Field Measurements for Forest Carbon Monitoring

Coeli M. Hoover
Editor

Field Measurements for Forest Carbon Monitoring

A Landscape-Scale Approach

Springer

Editor
Coeli M. Hoover
US Forest Service
Durham, NH
USA

ISBN: 978-1-4020-8505-5 e-ISBN: 978-1-4020-8506-2
DOI: 10.1007/ 978-1-4020-8506-2

Library of Congress Control Number: 2008931346

Cover Images
Main cover photo: Landscape-scale forest carbon monitoring site, Bartlett Experimental Forest, Bartlett, New Hampshire, USA. Photographer: Coeli Hoover

First inset photo: Meteorological/flux tower instrument shed and base of tower, Bartlett Experimental Forest, Bartlett, NH, USA. Photographer: Coeli Hoover

Second inset photo: Measuring the diameter of a downed log to quantify the amount of carbon stored in dead woody material at the Glacier Lakes Ecosystem Experiment Station in southern WY, USA. Photographer: John Bradford

Third inset photo: Technician climbing the meteorological/flux tower to perform maintenance. Bartlett Experimental Forest, Bartlett, NH USA. Photographer: Coeli Hoover

Fourth inset photo: Measuring soil respiration rates at the Niwot Ridge Ameriflux site near Nederland, CO, USA. Photographer: John Bradford

Printed on acid-free paper

9 8 7 6 5 4 3 2 1

springer.com

Preface

In the summer of 2003, a workshop was held in Portsmouth, NH, to discuss land measurement techniques for the North American Carbon Program. Over 40 scientists representing government agencies, academia and nonprofit research organizations located in Canada, the US and Mexico participated. During the course of the workshop a number of topics were discussed, with an emphasis on the following:

- The need for an intermediate tier of carbon measurements. This level of study would be more extensive than state-level inventories of the US Forest Service Forest Inventory and Analysis Program, but less detailed than intensive ecosystem studies sites such as those in Long Term Ecological Research network. This tier would ideally provide a basis to link and scale remote sensing measurements and inventory data, and supply data required to parameterize existing models (see Wofsy and Harriss 2002, Denning et al. 2005).
- The design criteria that such a network of sites should meet. The network and sampling design should be standardized, but flexible enough to be applied across North America. The design also needs to be efficient enough to be implemented without the need for large field crews, yet robust enough to provide useful information. Finally, the spatial scale must permit easy linkage to remotely sensed data.
- The key variables that should be measured at each site, and the frequency of measurement. Considerations include contribution of the process or property to quantifying and understanding carbon cycling at the site, usefulness as a model input, and practicality of implementation.

During the workshop, the participants developed a list of candidate variables for measurement at landscape-scale carbon monitoring sites, with recommended measurement intervals. The list was compiled and participants were asked to rate each variable as high, medium, or low priority. Numerical scores were assigned to each importance category and the list was compiled. The entire variable list, in order from highest score to lowest, is given in Table 1. General comments on the variable and suggested measurement frequencies (as discussed during the workshop) are also included. Through further discussion the list was refined. Some variables were included with others, for instance, geographic coordinates are part of the site description section. Some, such as nitrous oxide flux, were not included due to a

Table 1 Variables in order of numerical ranking, highest to lowest. Rows enclosed by brackets had tied rankings

Variable	Comments	Frequency
Aboveground biomass		Annually + mortality
Coarse woody debris	By size class	Every 5 years
Litterfall		Monthly or biweekly
Site history	Land use; disturbance	Once; annual update
Foliar nitrogen		Annual; seasonal
Forest floor mass		Varies with site age
Temperature		Continuous
Stand age (site age)	Mixed age stands?	Update every 5 years
Soil carbon dioxide (CO_2) flux		Monthly; biweekly
Precipitation		Continuous
{Coarse root biomass	Live and dead	Improve allometric equations
{Photon flux density	Paired – below canopy, clearing	Continuous
Vegetation height	LIDAR application	Every 5 years
Dead wood decomposition	From known event?	Annual
Leaf area index (LAI)	Coordinate with canopy height	Annual or monthly; seasonal
Soil moisture		Hourly or daily
{Woody shrubs	Mass/area of crown vs. height	Every 5 years
{Soil carbon: nitrogen ratio		Once
{Carbon 14 soil incubation	Expensive	Repeat every 5–10 years
Soil carbon stocks	To what depth?	Every 10 years
Net radiation		Continuous
Soil temperature		Hourly; daily
Branchfall	Include in litterfall section?	Varies by season
Specific gravity by species	Dead wood	Once; subsampled
Fine root biomass	Not practical to measure turnover	
Vapor pressure		Continuous
Litter decomposition		Once
% Canopy cover		Annual
Wood utilization	Where harvests occur	
Methane (CH_4) fluxes	Appropriate in forested wetlands	Weekly in wetlands
Dissolved organic carbon	Important in some ecosystems	Monthly or weekly
Nitrous oxide (N_2O) flux	If forest fertilization is used	Daily
Suspended sediment	How to relate to sample plots?	Monthly
Geographic coordinates	Part of site description	
Labile carbon		
Root allometry		
Litter moisture		

decision to focus on carbon. Other measurements were omitted for reasons of high cost, low priority, or limited applicability. The end result was a slightly smaller list of variables thought to capture the properties and processes of carbon cycling aboveground and belowground that would also provide the data necessary to link to remotely sensed information as well as existing process models.

Once the list was refined, workshop participants indicated the areas in which they had expertise. One of the desired outcomes from the workshop was a field manual focused specifically on landscape-scale carbon inventory and monitoring. In view of this objective, the variables were grouped into subsets, and writing teams were self-selected from the list of individuals with expertise in each area. The writing teams each had responsibility for a chapter of the manual; chapters covered one or more of the variables in Table 1. Each chapter team discussed available methods and recommended one; where alternate approaches may be necessary or appropriate those are included as well. In addition, for some variables, several levels of measurements may be described. In these cases there is a fundamental level for required implementation followed by one or more additional levels of measurement that may be applied if budgets allow or if site investigators require more detail. For the most comprehensive dataset, it is recommended that data on all of the variables included in the manual be taken. It is worth noting that while the suite of variables was chosen with the needs of flux modeling in mind, a flux tower is not a necessary component of a landscape-scale carbon monitoring installation.

This manual is intended to serve as a reference for scientists wishing to implement landscape-scale carbon measurements. As such, the level of detail is intermediate in nature; descriptions of procedures are not exhaustive, but include many references to supporting material. Each chapter generally includes background information, recommended methods, equipment needs, required calculations, and practical considerations; chapters vary in length and content since these decisions were left to the writing teams. Each chapter team has extensive field experience in their measurement area.

A smaller team met in Durham, NH, to discuss plot layout and design for the proposed network. Design considerations included: compatibility with existing nationwide forest inventory protocols in North America, practicality of installation in a variety of landscape types, suitability for conducting a wide range of measurements, and integration with measurements taken at other tiers in the North American Carbon Plan hierarchy. Chapter One provides a complete description of the rationale behind the landscape scale measurement tier as well as the plot design that was chosen during the layout and design meeting.

The opening part of the manual continues with Chapter Two, which outlines the information required for an adequate site description, and Chapter Three, Meteorological Measurements. From that point, the recommended landscape-scale carbon measurements are organized into the following parts: Aboveground Carbon Pools, Aboveground Carbon Fluxes, Belowground Carbon Pools and Fluxes, and Supplemental Variables for Carbon Cycle Modeling. Finally, a concluding part includes chapters on integrating field measurements with remote sensing data and flux measurements, and lessons learned from implementing the measurements at pilot sites with very different characteristics.

While this manual provides specific guidance on the installation of landscape-scale forest carbon monitoring sites in the framework of the North American Carbon Plan, it can also be used as a reference for those interested in various aspects of forest carbon inventory and monitoring. We expect that instruments and methods will continue to improve, but the fundamental techniques outlined in this volume should serve investigators well into the future.

Coeli M. Hoover
US Forest Service
Northern Research Station
Durham, NH 03824

Literature Cited

Denning S, Oren R, McGuire D, et al. (2005) Science Implementation Strategy for the North American Carbon Program. Report of the NACP Implementation Strategy Group of the U.S. Carbon Cycle Interagency Working Group. Washington, DC: U.S. Carbon Cycle Science Program. http://www.carboncyclescience.gov/docs.php

Wofsy SC, Harriss RC (2002) The North American Carbon Program (NACP). Report of the NACP Committee of the U.S. Interagency Carbon Cycle Science Program. Washington, DC: US Global Change Research Program. http://www.carboncyclescience.gov/docs.php

Disclaimer

The use of trade, firm, or corporation names in this publication is for the information and convenience of the reader. Such use does not constitute an official endorsement or approval by the US Department of Agriculture or the Forest Service of any product or service to the exclusion of others that may be suitable.

Acknowledgements

This book is the result of a collaborative effort by many individuals. The initial workshop was organized by Research Project 4104 of the US Forest Service Northern Research Station and the University of New Hampshire, with support from the National Institute of Global Environmental Change and the US Forest Service Northern Global Change Program. Workshop co-conveners were Linda Heath and David Hollinger, US Forest Service, and John Aber, University of New Hampshire. Sue McGrane provided extensive logistical support during all phases of the workshop. Mike Nichols compiled all workshop presentations on CD for distribution to participants.

The pilot studies of carbon monitoring methods described in this volume were funded by the US Forest Service Northern Global Change Program and NASA grant CARBON/04-0225-0191.

The individual chapters benefited from thoughtful reviews from a number of individuals whose time is greatly appreciated: Pierre Bernier, Johnny Boggs, John Bradford, Ken Clark, Peter Curtis, Mats Fröberg, Mike Gavazzi, Christian Giardina, Paul Hanson, Linda Heath, Randy Kolka, Harry McCaughey, Linda Pardo, Chris Potter, Neal Scott, M. L. Smith, Sara Strickland, and Peter Weishampel. A special thanks to our authors who took the time to review colleagues' chapters in addition to their own writing duties.

Richard Birdsey facilitated pilot studies, organized follow-up meetings, reviewed the volume for content and consistency, and provided support to move the project along. Linda Heath provided valuable feedback and assistance many times during the process of compiling this volume. Susan Stout has been extremely patient regarding the time required by the project, and Sue McGrane deserves a great deal of thanks for the many tasks that she has carried out during the process of taking the idea of a landscape-scale forest carbon monitoring program from a workshop to a book.

This book has not been subject to policy review by the US Forest Service or any other US Government sponsors and, therefore, does not represent the policies of any agency.

Contents

Contributors

Jeanne Anderson
Complex Systems Research Center, University of New Hampshire,
39 College Road, Durham, NH 03824

Pierre Bernier
Natural Resources Canada, Canadian Forest Service,
PO Box 10380, Stn. Sainte-Foy, Quebec G1V 4C7

Richard A. Birdsey
US Forest Service, Northern Research Station, 11 Campus Blvd., Suite 200,
Newtown Square, PA 19073

John B. Bradford
US Forest Service, Northern Research Station, 1831 Hwy 169 E. Grand Rapids,
MN 55744

Sandra Brown
Winrock International, Ecosystem Services Unit, 1621 N. Kent Street, Suite 1200,
Arlington, Virginia 22209

Andrew J. Burton
School of Forest Resources and Environmental Science, 1400 Townsend Drive,
Michigan Technological University, Houghton, MI 49931

David C. Chojnacky
US Forest Service, Washington, DC
Current address: Virginia Tech Department of Forestry, 144 Rees Place, Falls
Church, VA 22046

Peter S. Curtis
Department of Evolution, Ecology, and Organismal Biology, The Ohio State
University, 318 W. 12th Avenue, Columbus, OH 43210

Kenneth J. Davis
Department of Meteorology, The Pennsylvania State University, 512 Walker Building, University Park, PA 16802

Mark J. Ducey
Department of Natural Resources, University of New Hampshire, 215 James Hall, Durham, NH 03824

Matthew Fladeland
NASA Ames Research Center, Mail Stop 242-2, Moffett Field, CA 94035

Mats Fröberg
Department of Energy, Oak Ridge National Laboratory, Oak Ridge, TN,
Current address: Department of Forest Soils, Swedish University of Agricultural Sciences, PO Box 7001, SE 750 07 Uppsala Sweden

Christian P. Giardina
US Forest Service, Pacific Southwest Research Station, 60 Nowelo Street, Hilo, HI 96720

Jeffrey H. Gove
US Forest Service, Northern Research Station, 271 Mast Rd, Durham, NH 03824

Timothy G. Gregoire
School of Forestry and Environmental Studies, Yale University, 360 Prospect St, New Haven, CT 06511

Paul J. Hanson
Environmental Sciences Division, Oak Ridge National Laboratory,
Bethel Valley Road, Building 1062, Oak Ridge, TN 37831

David Y. Hollinger
US Forest Service, Northern Research Station, 271 Mast Road, Durham, NH 03824

Noah J. Karberg
US Forest Service, Northern Research Station, 410 MacInnes Drive, Houghton, MI 49931

Randall Kolka
US Forest Service, Northern Research Station, 1831 Highway 169 E., Grand Rapids, MN 55744

Mikaila Milton
National Science Foundation, Arlington, VA
Current address: National Park Service, National Capital Parks-East, 1900 Anacostia Drive S.E., Washington, DC 20020

Scott Ollinger
Complex Systems Research Center, University of New Hampshire, 56 College Road, Durham, NH 03824

Chris Potter
NASA Ames Research Center, Mail Stop 242-2, Moffett Field, CA 94035

Kurt S. Pregitzer
Department of Natural Resources and Environmental Science, University of Nevada, 1000 Valley Road, Reno, Nevada 89512

Michael G. Ryan
US Forest Service, Rocky Mountain Research Station, 240 W. Prospect Ave., Fort Collins, CO 80526
Affiliate Faculty, Graduate Degree Program in Ecology, Colorado State University, Fort Collins, CO

Neal A. Scott
Woods Hole Research Center, Falmouth, MA
Current address: Department of Geography, Queen's University, D201 Mackintosh-Corry Hall, Kingston, Ontario K7L 3N6

Marie-Louise Smith
US Forest Service, Northern Research Station, Durham, NH
Current address: US Forest Service, Legislative Affairs, 201 14th Street, SW, Washington, DC 20250-1130

Harry T. Valentine
US Forest Service, Northern Research Station, 271 Mast Rd, Durham, NH 03824

Peter Weishampel
Department of Soil, Water and Climate, University of Minnesota, 1991 Upper Buford Circle, St. Paul, MN 55108

Michael S. Williams
Food Safety Inspection Service, 2150 Centre Ave, Building D, Fort Collins, CO 80526

Part I
Establishing a Landscape-Scale Forest Carbon Monitoring Site

Chapter 1
Defining a Landscape-Scale Monitoring Tier for the North American Carbon Program

David Y. Hollinger

Abstract Better knowledge of carbon stocks and fluxes is needed to understand the current state of the carbon cycle and how it might evolve with changing land uses and climatic conditions. For Canada, the United States, and Mexico, the North American Carbon Program (NACP) has been devised to measure and understand the sources and sinks of CO_2, CH_4, and CO in North America and adjacent ocean regions. At one scale, there are presently several national networks, each containing many thousands of plots, that provide broad characterization of carbon stocks in different land types (especially forest). At a much finer scale and far fewer are intensive sites where detailed and frequent measurements of various carbon cycle stocks and processes are made. There exists a large discontinuity in spatial coverage, frequency of sampling, and scope of research between the intensive sites and the extensive inventory plots. We define here an intermediate (~1 km^2), landscape-scale sampling system to help bridge the gap between these different measurement intensities.

Keywords Carbon storage, landscape, NACP, plot design, sampling

1.1 Introduction

The North American Carbon Program (NACP) is an interdisciplinary plan for research on the carbon cycle (Wofsy and Harriss 2002). The objective of the NACP is to measure and understand the sources and sinks of CO_2, CH_4, and CO in North America and adjacent ocean regions. Of particular concern is developing knowledge sufficient to predict the future state of the carbon cycle and to learn how to effectively manage long-lived carbon sinks (CCSP 2007). The NACP is designed

D.Y. Hollinger
US Forest Service, Northern Research Station, 271 Mast Road, Durham, NH 03824
E-mail: dhollinger@fs.fed.us

C.M. Hoover (ed.) *Field Measurements for Forest Carbon Monitoring*,

to generate new techniques and models that will provide long term, mechanistically detailed, spatially and temporally resolved carbon fluxes across North America.

The NACP is distinct from previous studies of the carbon cycle in two principal ways: (i) by nature of its integrative, continental-scale approach that builds on a range of different and complementary observations, and (ii) by using powerful bio-physical models to extend information from one domain to another. The scientific strategy of the NACP is to link these different observational scales and methods with new measurements at intermediate scales. Also critical to the success of the NACP is the integration of observations and models using advanced data-model fusion techniques (Wofsy and Harriss 2002). The purpose of this manual is to describe the comprehensive set of measurements to be obtained from these intermediate-scale sites that will help achieve the aims of the NACP.

A principal element of the NACP is long-term atmospheric measurements of the major carbon gases: CO_2, CH_4, and CO. Three-dimensional concentration fields of these gases will be combined with their anthropogenic source distributions using transport models in inverse ("top down") analyses to infer regional carbon sources and sinks (Denning et al. 2005). These measurements will be carried out from a variety of platforms, including networks of tall towers, aircraft, and eventually, new satellite-borne sensors. Data obtained from sensors mounted on tall towers will be recorded continuously, while aircraft and satellite measurements will most likely be made on weekly or monthly intervals to enable the following of seasonal variations in C exchange. An important part of the atmospheric measurement program will be regional-scale field experiments to test out procedures and coordination among research teams.

A second important element of the NACP is land surface carbon (C) inventories, fluxes, and process studies. In this instance, the change in C stocks, or direct measurements of land-atmosphere gaseous C fluxes, will be scaled-up to the region ("bottom up") in a variety of modeling frameworks. These frameworks include remote sensing, biophysical, and statistical paradigms. New process-level studies are needed to identify critical mechanisms that regulate C exchange. Complementing land surface measurements and modeling will be similar efforts over the oceans adjacent to North America.

1.2 A Hierarchical Approach for Land Inventories and Processes Studies

The NACP proposes to combine enhanced carbon inventories, remote sensing, and models to provide a complete carbon accounting for the land sector and comprehensive analysis of the mechanisms driving the fluxes. New emphases on carbon accounting, on lands (peatlands, scrub land, suburban landscapes) and carbon pools (roots, shrubs) that lack comprehensive inventories, and on scaling with remote sensing and models, will greatly improve the carbon budget for North America and more localized regions.

Table 1.1 Multi-tier concept with likely variables central to a land observation system

Example data elements	Intensive sites (e.g., LTER, AmeriFlux, NEON)	Landscape-scale sample (new)	Extensive inventory (FIA and NFI)	Mapping and remote sensing
Land cover class	X	X	X	X
Leaf area index	X	X	X	X
Live biomass	X	X	X	X
Land cover change		X	X	X
Wildfire disturbance	X	X	X	X
Climate variability	X	X		
Soil CO_2 flux	X	X		
Methane flux	X	X		
Dissolved organic C	X			
Ecosystem CO_2 flux	X			

A hierarchical approach to these land studies will support multi-scale interpretations, with intensive studies providing access to details and mechanisms that are extended using remote sensing, extensive inventories, and mechanistic models. This multi-scale approach to land data will join the atmospheric and ocean studies as components in a unified analysis framework. Constraints from the atmosphere and oceans will increase the sophistication and accuracy of the estimates based on land data.

Much of the infrastructure for an NACP sampling hierarchy is already in place. At one end of the hierarchy is complete spatial coverage of North America for a few remotely sensed properties of the land surface, and at the other extreme are detailed carbon cycle process studies taking place at a limited number of long-term research sites (Table 1.1). In general, precision of carbon stock estimates and sophistication of understanding of carbon cycle processes increases for a decreasing spatial domain. We discuss below the three hierarchical levels of information presently available that will be essential components of NACP success. In considering the characteristics of these levels and their carbon cycle information, the authors of the NACP and NACP implementation plan (Wofsy and Harriss 2002, Denning et al. 2005) realized that there is a gap in the hierarchy that will limit the ability of land-based data to contribute to NACP goals. This manual describes a new research echelon, an intermediate level at the landscape scale of the hierarchy, which will fill this gap and help ensure successful completion of the NACP.

1.3 Present Levels in a Land-Based Carbon Cycle Hierarchy

1.3.1 Extensive Coverage with Remote Sensing

The NACP goal of providing spatially resolved, continental-scale information about carbon sources and sinks requires comprehensive information about the land surface that can only be provided by remote sensing. Needs include land classification by vegetation type, land-cover change detection, and estimates of land surface parameters that can be related to C uptake and loss.

Current US land inventory systems use a combination of high-altitude aerial photography and Landsat Thematic Mapper (TM) data to sample the largest scale and to detect change. TM data have been available since 1982 on a 16-day repeat cycle with a spatial resolution of 30 m, and since 1972 at a 72 m resolution.

Several sensors with coarse spatial resolution but higher (daily) temporal resolution have been used to drive terrestrial biogeochemistry models that estimate carbon stocks and fluxes, typically with limited information on land use history. The NOAA AVHRR (Advanced Very-High-Resolution Radiometer) sensors have provided fairly continuous global daily coverage at a nadir resolution of 1.1 km since 1979, but provide only limited spectral information. SeaWiFS provided a vegetation index of higher quality but for a shorter time period (Behrenfeld et al. 2001). The NASA MODIS sensor, launched on the Terra spacecraft in 1999, and Aqua in 2002, provide higher spectral resolution data (36 bands) at a moderate resolution (500–1,000 m, depending on wavelength). A variety of data products are available, including vegetation indices, land cover and cover change, surface reflectance, surface temperature, presence of fire, leaf area index (LAI), fraction of photosynthetically active radiation (FPAR) absorbed by the canopy, and terrestrial net primary productivity (NPP).

Additional sensors in the planning stage would provide new and useful information, including aspects of canopy chemistry and structure and soil moisture. Light detection and ranging (LIDAR) technology is particularly promising because satellite monitoring of vegetation structure has been beyond the scope of previous sensors, and these measurements are closely related to above-ground biomass, one of the most dynamic ecosystem carbon pools.

Several specific needs have been identified to provide data to the NACP on above-ground carbon stocks using remote sensing products and *in situ* data, including: (1) timely systematic and routine processing of satellite data from the North American continent into land cover and land cover change products, including both natural and human disturbances; (2) integration of satellite observations with *in situ* measurements of carbon stocks and existing inventories; (3) augmentation of satellite and *in situ* estimates of carbon stocks with airborne and surface measurements; and (4) development of appropriate estimation models for carbon stocks both above- and below-ground.

1.3.2 Extensive Sampling with Inventory Techniques

Across much of North America, large-scale land inventories are conducted by the USDA Forest Service (the Forest Inventory and Analysis, or FIA), Natural Resources Conservation Service (National Resources Inventory, or NRI), and the Canadian provincial/territorial governments and Federal government (National

Forest Inventory or NFI), employing multi-tier sampling strategies using remote sensing and ground measurements. These continuous inventories provide baseline information about land cover, management intensity, productivity, and disturbance that can be used to estimate carbon stock changes over 5–10 year periods. Very high sampling intensity (within continental US forestland, for example, there is an average of one sampling point every 24 km^2) allows detailed description of some of the causes of observed carbon stock changes, such as the effects of vegetation growth, mortality, and harvesting. Historical data are available to trace land use and management history.

Where available, land inventories provide a representative, statistically based sample of ecosystems and their range of conditions in response to past disturbances. Current land inventories are limited by incomplete data from certain regions and ecosystem types, lack of complete ecosystem C measurements, limited temporal resolution, and lack of easily available and usable historical data. In the US, the FIA program does an excellent job with its coverage of all forested lands, regardless of ownership, employing standardized sampling on a uniform grid. Data are not yet available from all plots, especially interior Alaska, and unfortunately there is no FIA equivalent for non-forest land, especially grazing land in the West. Major portions of Canada and Mexico have been at best sparsely sampled. Forest inventories have historically excluded urban and suburban areas, but these gaps are gradually being addressed.

In the US FIA program, additional carbon pools consisting of downed woody material, the forest floor, and 8 in. of mineral soil are sampled in a 1/16th subset of the FIA plots (the Phase 3 plots). Soil sampling consists of bulk density, carbon content, litter and duff depth, and depth to a restrictive layer below the mineral soil. Down woody debris sampling is transect-based and separates coarse and fine woody debris (see the FIA Phase 3 Field Guide, sections 11 and 14 for more information). Stumps, live and dead roots, and deeper mineral soils are presently not sampled.

In the US the NRI has data on land use and natural resource conditions and trends for non-Federal lands recorded in surveys at roughly 5-year intervals from 1982 (Nusser and Goebel 1997). Much of this information, including the STATSGO2 and SSURGO soil databases, is available through the USDA Geospatial Data Gateway (http://datagateway.nrcs.usda.gov/).

In the past, land inventories were designed to provide periodic estimates with a temporal resolution of 5–15 years, sufficient for some applications, but incompatible with the temporal resolution needed by the NACP. Newly implemented designs for forest inventories address temporal resolution using successive sample "panels" to approximate continuous sampling. Each panel is re-sampled with a period of 5–10 years. Supplemental data with higher time resolution can be merged with the inventory data to estimate the major causes of variations in C flux; productivity, mortality, harvest, and land use change. Sources of supplemental data include flux towers (productivity and trace gas dynamics), aerial and satellite disturbance surveys (land use change, damage and mortality), and timber and agricultural product surveys (harvest quantities).

1.3.3 Intensive Measurement Sites

Intensive sites provide direct estimates of C flux and C stock changes across a range of temporal scales. In addition, research at these sites includes detailed studies of the mechanisms controlling carbon fluxes. Data from intensive sites will be critical for developing and testing models, for interpreting large-scale patterns, and for constraining models.

There are perhaps 50–100 sites carrying out intensive studies of the carbon cycle across North America. Many of these sites are part of the Fluxnet Canada, AmeriFlux or emerging NEON networks, and make continuous measurements of land-atmosphere CO_2 exchange from meteorological research towers. Summed over the course of a day, month, season, or year, data from these sites provide direct measures of ecosystem CO_2 source or sink strengths. Most intensive sites provide information specific to one ecosystem type. Because of the relatively small number of intensive sites, the existing range of conditions following past disturbances within most ecosystem types is not fully represented.

Data from intensive sites help estimate parameter values or test physiological models of C exchange and are critical to relating fluxes and remote-sensing data. Companion physiological and ecological measurements enable the partitioning of carbon fluxes into plant and soil components and can reveal mechanisms responsible for variation in these fluxes. At several sites, biomass-based estimates of C storage have validated C budgets from direct flux data (e.g., Barford et al. 2001, Curtis et al. 2002).

1.4 A New, Intermediate Level of Study is Needed: Process and Stock Monitoring at Landscape-Scale Sites

There is a large discontinuity in spatial coverage, frequency of sampling, and scope of research between the intensive sites and the extensive inventory plots. There are probably 1,000-fold more inventory plots than intensive sites in North America. For most of the US, Canada, and parts of Mexico, inventory plots provide excellent coverage of the full range of ecosystem states and with abundant replication. However, inventory plots are typically visited only once in 5–10 years, whereas intensive sites provide some data at intervals of 30 min, allowing a better understanding of the mechanisms underlying the temporal aspects of carbon exchange. Inventory sites provide information about the major C stocks (stems, soil, coarse woody debris) but typically do not assess other important C pools including foliage and roots.

To provide better representation of the range of ecosystem conditions with detailed, high-resolution measurements of ecosystem C, and to facilitate linking intensive sites with remote sensing, a new set of sites intermediate in intensity of measurement and number is needed. It is not practical to work intensively in all ecosystem types and across the full range of land use histories in North America. Yet, process data from across this range is crucial for robust analysis. A new network of medium-intensity sites at the landscape-scale is needed to achieve the goals of the NACP.

The purpose of this manual is to describe the comprehensive set of measurements to be obtained from these sites that will help achieve the aims of the NACP. The measurements made at these sites must meet several requirements. First, they should provide valid statistical estimates of major C stocks and their spatial distribution, and be able to detect change in these stocks over a short time period (one to a few years). Secondly, they should be representative of a large enough area so that surface characteristics can be directly detected by the full range of Earth observing satellites and used in modeling studies. Thirdly, they should provide the terrestrial information necessary to parameterize and run vegetation-dynamics and biogeochemically-based models of ecosystem function such as TEM, Biome-BGC, CENTURY, CASA, and IBIS.

Measurements at these sites will address major components and processes in the carbon balance, including annual net primary productivity (NPP), leaf area index, leaf nutrients, soil respiration, litterfall, dynamics of coarse woody debris, and CH_4 flux. Depending on technology development, it may soon be practical to measure C balance with eddy flux at many of these sites. It will be important to obtain accurate records of land use, including past history as well as current management practice. Environmental conditions, including soil moisture and solar radiation (see Chapter 3), should be measured at each site to facilitate assessment with ecosystem models.

The new sites will serve as vital links between the approximately 100 intensive sites and the much larger number of inventory sites. Because these landscape-scale sites will measure directly components of C balance, they will be a key resource for testing models developed at the intensive sites. Direct measures of components of the carbon balance will also be critical for setting appropriate conversion factors for the quantities that need to be estimated in inventories.

Coverage of North American ecosystems by intensive sites that make up the LTER and AmeriFlux networks is incomplete (Hargrove et al. 2003) and will remain sparse even with the additional intensive sites of the National Ecological Observatory Network (NEON). In addition, the eddy covariance methodology used at many intensive sites is poorly suited to steep topography, so intensive carbon cycle information is often not available, increasing the value of the landscape-scale sites in these situations. Similarly, these landscape-scale sites will have extra value in regions and vegetation types lacking inventory plots.

1.5 Overview of a Landscape-Scale Site

The first goal of the new sampling system is to provide statistical estimates of major C stocks and fluxes across an area of land. The new system must therefore be compatible with existing inventory systems to provide ready comparison and extrapolation across the landscape. This requirement suggests the use of FIA-compatible plots within the United States and NFI-compatible plots within Canada.

The second goal of the new system is to be able to tie information on C stocks and fluxes to remotely sensed variables over the same landscape. The spatial resolution (pixel size) of some important land resource satellites is large (up to 1 × 1 km), suggesting that a relatively large land area is necessary for a medium-intensity site. A third goal for these sites is to provide measurements of carbon storage or loss that can be compared with and are independent of other measurements based on different methodologies such as eddy flux towers. This suggests that some intermediate intensity sites may wish to incorporate eddy flux measurements or other "continuous" measurements such as soil CO_2 flux. Over a forest, the flux source ("footprint") region of eddy covariance measurements may extend hundreds of meters, again suggesting a relatively large scale.

To meet these separate goals, the standard landscape-scale site will consist of inventory plots arrayed across a minimum 1 × 1 km land area (Fig. 1.1). The gridded area should be centered on a meteorological tower or other special research installation. Use of a sampling grid is a relatively efficient and unbiased way to select sampling locations (Haining 1993). Between 16 and 36 inventory plots may be adequate

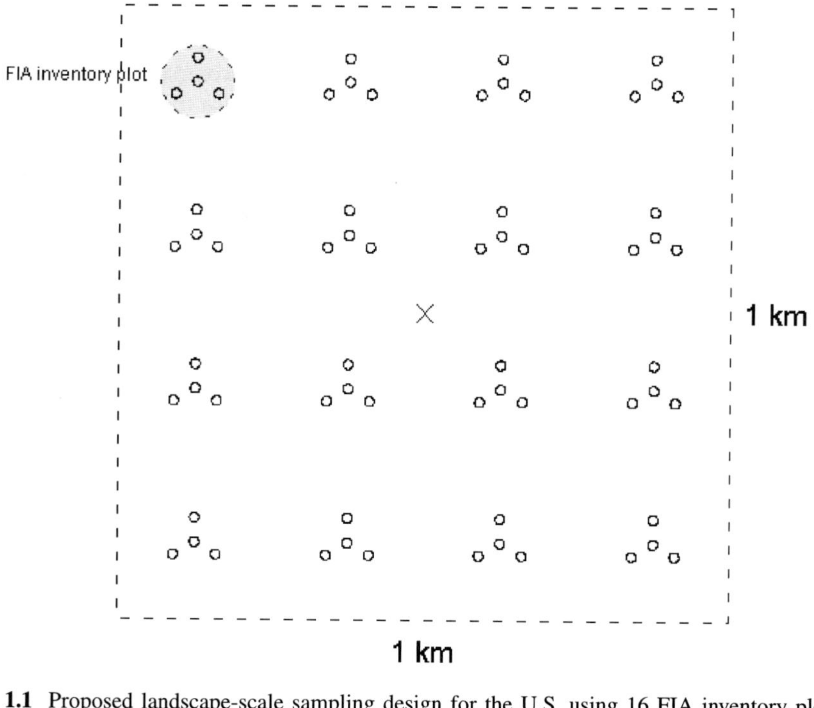

Fig. 1.1 Proposed landscape-scale sampling design for the U.S. using 16 FIA inventory plots. Researchers may need to install additional plots to ensure that measurement precision is within desired limits. The FIA design is shown in more detail in Fig. 1.3. The cross represents a meteorological or flux tower

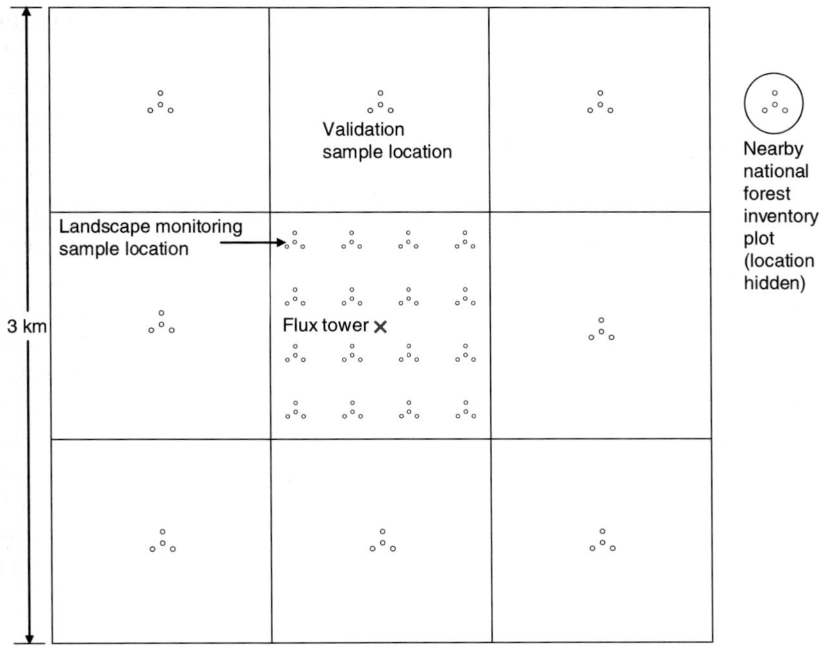

Fig. 1.2 Landscape-scale sampling area expanded to 3 × 3 km to include additional landscape features. The individual 1 × 1 km squares may be arranged in different configurations depending on the features of the landscape that are targeted for sampling. In this example, the additional 1 × 1 km squares are areas where LIDAR imagery will be acquired, and only a very small number of additional field sampling locations established to validate estimates

depending upon anticipated variability of the land area and desired precision. The 1 × 1 km land area may be expanded by adding adjacent 1 × 1 km sampling areas if estimates for a larger variety of landscape conditions are required (Fig. 1.2).

Other measures such as soil characteristics, soil respiration, litterfall, canopy N%, etc. should be carried out at or immediately adjacent to each inventory plot (note that in the FIA system, four of the grouped circular subplots constitute a "plot"). If it is not feasible to sample at each inventory plot (because of high cost, such as for ^{14}C analysis), groups of plots may be combined and one sample taken for each group. The goal should be to sample extensively across the 1 × 1 km grid rather than intensively at one or several inventory plots. This design should provide an unbiased estimate of carbon stocks and fluxes across the 1 km "pixel" but this may represent only one or a few actual satellite pixels (probably less due to registration issues). To improve the connection with satellite data, multiple 1 × 1 km sites may be needed.

1.5.1 Inventory Plots

The United States FIA and Canadian NFI forest inventory plots differ in design, and, in turn, both differ from a system proposed for the Mexican states of Jalisco and Colima (Wulder et al. 2001, USDA 2007). In the US, all FIA units have implemented a common sampling design consisting of four clustered, 24 ft radius subplots (total sampled area of 672 m^2 or 0.166 acre) for trees at least 5 in. in diameter and four 6.8-ft radius microplots (total area of 54 m^2 or 0.0133 acre) for smaller trees (Fig. 1.3). Subplot 1 is the center of the cluster with the other three subplots located 120 ft away at azimuths of 360°, 120°, and 240°, respectively.

In contrast, the NFI plots are larger and laid out concentrically (Anonymous 1999). The "large tree plot" has a fixed-radius of 11.28 m (37.01 ft) and area of 400 m^2 (1/10 acre), and the "small tree plot" has a fixed-radius of 3.99 m (13.09 ft) with an area of 50 m^2 (0.0124 acre).

The Mexican system (Anonymous 2002) consists of a 30 x 30 m square primary sampling unit (PSU) containing five clustered subplots, each of 5 m radius (total sample area of 392.7 m^2 or 0.097 acre). The PSU is divided into a grid of nine 10 × 10 squares with subplots centered in the corner and middle grid cells of the PSU. Trees larger than 12.5 cm DBH are measured on each subplot and saplings (2.5 cm ≤ DBH ≤ 12.5 cm) are measured on 3 m radius plots centered on the subplots (total area of 141.4 m^2 or 0.035 acre). Each of these sampling designs serves as a satisfactory solution for assessing tree and shrub carbon stocks on a plot of land. To maintain compatibility with existing inventory systems, we recommend that inventory plots match the design of the standard inventory plot in their country. We discuss

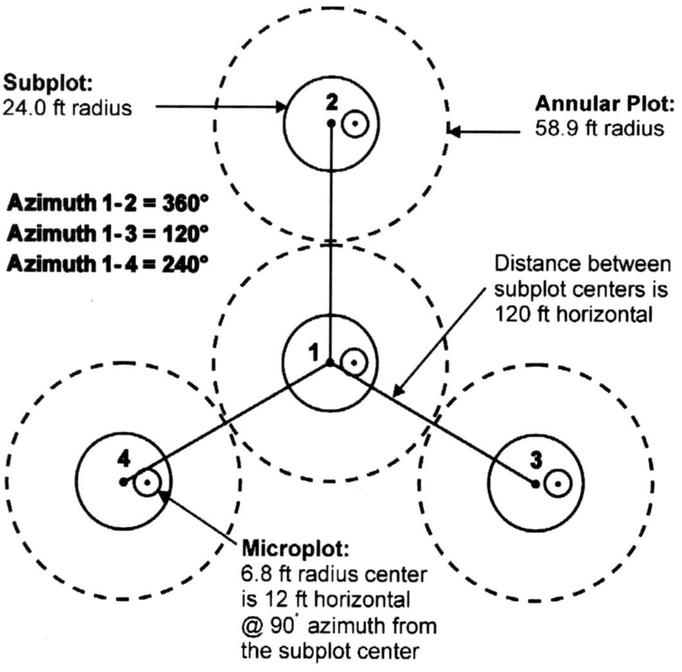

Fig. 1.3 USDA Forest Service Forest Inventory and Analysis plot. The "annular" (58.9 ft radius) plots are not used in the landscape-scale sampling design

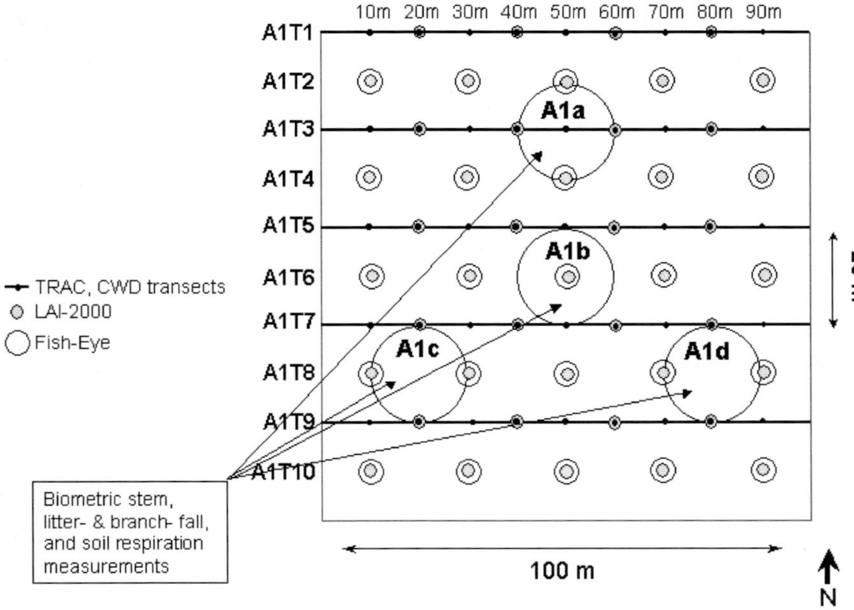

Fig. 1.4 Detail of additional measurement locations around a cluster of FIA-type subplots. These measurements would be repeated around each cluster of subplots

exceptions to this rule later in the manual. One possible exception relates to the assessment of coarse woody debris (CWD). The US, Canadian, and proposed Mexican systems use "Y"-shaped, "cross", and straight line-intersect transects, respectively, for measuring CWD. Recent work suggests that "Y"-shaped or "cross" designs lead to estimation errors (Gregoire and Valentine 2003) and that a single transect with length equivalent to the sum of the segments in the current systems would be better. In this manual, however, we recommend an improved method for sampling CWD, perpendicular distance sampling (Chapter 6).

For some measurements the sampling intensity may need to be increased, necessitating an overlay on the plot design. An example of this (Fig. 1.4) shows one FIA cluster of plots overlain with additional sampling locations for leaf area index, litter fall, soil respiration and other factors. Additional information is included in the chapters describing specific measurements.

1.5.2 Precision of Estimates

Since the inventory plots provide a statistical sample of the carbon pools across the landscape-scale site, it is important to declare up front what would constitute an acceptable level of error in the estimates. Setting this acceptable error level too high may result in poor correlation with remotely sensed variables or have insufficient statistical power to reject a false null hypothesis. A reasonable, acceptable margin of error for studies of carbon pool size is plus or minus 10%. For estimating a mean with 95% confidence, the sample size n needed is:

$$n = \frac{16}{(d / \sigma)^2}$$

where d is the acceptable range of the error and σ is the standard deviation. As an example, Hollinger (unpublished) found that the standard deviation of biomass estimates made from FIA-sized subplots in a relatively homogeneous spruce-hemlock forest was 33% of the mean. To obtain a precision of ±10% of the mean with 95% confidence, 44 subplots (11 FIA plots) would be necessary. Over a more variable landscape, more plots would be necessary to meet this target or it would be necessary to accept a reduced precision. Designs with 16 FIA plots will reach the desired precision when $\sigma < 40\%$ and with 36 FIA plots when $\sigma < 60\%$ of the mean. The variance (σ^2) will depend upon the pool or flux being sampled, but should be estimated for all important parameters so that the necessary sample size can be calculated. Carbon in soils, for example, can be highly variable, necessitating additional sampling.

1.6 Implementation Issues

The new landscape-scale monitoring tier should be comprehensive in its coverage of North American ecosystems, and represent a variety of disturbances and stages of ecosystem recovery following disturbance. Each landscape-monitoring site should ideally be representative of landscape conditions over a larger area. Because some of the envisioned measurements require multiple site visits per year (as frequently as weekly for periods of high litterfall or for soil respiration measurements), it is efficient to use clusters of sampling locations as described here to minimize travel time between sampling locations, and to take advantage of existing infrastructure such as that found at established intensive sites.

The current network of intensive carbon monitoring sites should be considered a candidate list for enhancement with landscape-scale monitoring. Some of these sites already include landscape-scale carbon monitoring (for example, Ehman et al. 2002, Law et al. 2003). However, the current network of intensive monitoring sites does not fully represent all ecosystems and contains significant gaps in highly disturbed areas and steep terrain. Additional candidate sites to expand the landscape-scale network that already have some established infrastructure include experimental farms and forests maintained by government agencies, universities, and the private sector, and reserved areas that do not have restrictions on research activities. The analysis by Hargrove et al. (2003) is a good start at identifying areas that are under-represented, but this analysis should be expanded to reflect the needs of landscape-scale monitoring to include better representation of disturbed and managed areas.

Since data from landscape-scale sites are intended for use in a variety of applications under the NACP, standardized methodologies and outputs are required. However, in many cases the new measurements will be overlaid on existing intensive sites and the spatial distribution of landscape characteristics for sampling are highly variable, so some flexibility in application of the design principles described in this document is required. As mentioned earlier, the exact number of sampling locations required within a 1 × 1 km land area depends on homogeneity within the area. If multiple 1 × 1 km sampling areas are required, the spatial arrangement of the sampling areas may be adapted to the landscape. It is highly recommended that the individual sampling areas be no smaller than 1 × 1 km to allow co-registration with information from satellite sensors. Sampling protocols within the 1 × 1 km land areas may be adapted to take advantage of existing sample plots, or to stratify the landscape to increase the probability of randomly locating samples in important conditions that represent small proportions of the landscape. Finally, the exact suite of measurement may vary between landscape monitoring sites depending upon local conditions. When considering variations from the recommended design, however, the ability to provide standardized outputs that are compatible with those from other sites must be maintained. Thus, any local variation should be **in addition to** the standard design.

Other factors should also be considered in developing a sampling design and measurement protocol for a landscape-scale monitoring site under the NACP. Adherence to the NACP policies for data management (Cook and Thorton 2005) is required. Sampling locations should be accurately geo-referenced, and the coordinates made available to users without restrictions. This may require obtaining landowner permission to release the coordinates or other site data. Security of the site from unwanted manipulation, and of the installed equipment, should be a major consideration in site selection since a potentially long period of measurement and monitoring will be important to the NACP. And finally, safety must be the top priority for all measurement and monitoring activities.

Literature Cited

Anonymous (1999) A plot-based National Forest Inventory design for Canada. An interagency partnership project. NFI design version 2.0, March 31, 1999

Anonymous (2002) Proyecto piloto para el inventario y monitoreo de los recursos naturals. (Pilot project for inventorying and monitoring ecosystems resources.) Fideicomiso para la Administración del programa de desarrollo forestall del estado Jalisco. Guadalajara, Jalisco, México. December, 2002

Barford CC, Wofsy SC, Goulden ML, et al. (2001) Factors controlling long- and short-term sequestration of atmospheric CO2 in a mid-latitude forest. Science 294:1688–1691

Behrenfeld MJ, Randerson JT, McClain CR, Feldman GC, Los SO, Tucker CJ, Falkowski PG, Field CB, Frouin R, Esaias WE, Kolber DD, Pollack NH (2001) Biospheric primary production during an ENSO transition. Science 291:2594–2597

CCSP (2007) The First State of the Carbon Cycle Report (SOCCR): The North American Carbon Budget and Implications for the Global Carbon Cycle. A Report by the U.S. Climate Change

Science Program and the Subcommittee on Global Change Research. In: King AW, Dilling L, Zimmerman GP, Fairman DM, Houghton RA, Marland G, Rose AZ, Wilbanks TJ (eds.) National Oceanic and Atmospheric Administration, National Climatic Data Center, Asheville, NC, USA. 242 p. http://cdiac.ornl.gov/SOCCR/final.html

Cook RB, Thorton P (2005) Data management for the North American Carbon Program. Workshop Report. New Orleans, Louisiana. Jan. 25–27, 2005. 48 p http://www.carboncyclescience.gov/docs.php

Curtis PS, Hanson PJ, Bolstad P, Barford C, Randolph JC, Schmid HP, Wilson KB (2002) Biometric and eddy-covariance based estimates of annual carbon storage in five eastern North American deciduous forests. Agricultural and Forest Meteorology 113: 3–19

Denning S, Oren R, McGuire D, et al. (2005) Science Implementation Strategy for the North American Carbon Program. Report of the NACP Implementation Strategy Group of the U.S. Carbon Cycle Interagency Working Group. Washington, DC: U.S. Carbon Cycle Science Program. http://www.carboncyclescience.gov/docs.php

Ehman JL, Schmid HP, Grimmond CSB, Randolf JC, Hanson PJ, Wayson CA, Cropley FD (2002) An initial intercomparison of micrometeorological and ecological inventory estimates of carbon exchange in a mid-latitude deciduous forest. Global Change Biology 8(6):575–589

Gregoire TG, Valentine HT (2003) Line intersect sampling: Ell-shaped transects and multiple intersections. Environmental and Ecological Statistics 10:263–279

Haining R (1993) Spatial data analysis in the social and environmental sciences. Cambridge University Press, Cambridge. 431 p

Hargrove WW, Hoffman FM, Law BE (2003) New analysis reveals representativeness of the AmeriFlux network. EOS Transactions AGU 84(48):529

Law BE, Sun OJ, Campbell J, Van Tuyl S, Thornton PE (2003) Changes in carbon storage and fluxes in a chronosequence of ponderosa pine. Global Change Biology 9:510–524

Nusser SM, Goebel JJ (1997) The National Resources Inventory: A long-term multi-resource monitoring programme. Environmental and Ecological Statistics 4:181–204

U.S. Department of Agriculture, Forest Service (USDA) (2007) Forest inventory and analysis national core field guide, Volume 1: field data collection procedures for phase 2 plots, version 4.0. U.S. Department of Agriculture, Forest Service, Washington Office. Internal report. On file with: U.S. Department of Agriculture, Forest Service, Forest Inventory and Analysis, 201 14th St., Washington, D.C., 20250. http://www.fia.fs.fed.us/library/field-guides-methods-proc/

Wofsy SC, Harriss RC (2002) The North American Carbon Program (NACP). Report of the NACP Committee of the U.S. Interagency Carbon Cycle Science Program. Washington, DC: US Global Change Research Program. http://www.carboncyclescience.gov/docs.php

Wulder M, Gillis M, Luther J, Dyk A (2001) A Guide to the Estimation of Canada's National Forest Inventory Attributes from Landsat TM Data. Natural Resources Canada, Canadian Forest Service, Pacific Forestry Centre. Working paper. 84p

Chapter 2
Study Site Characterization

Chris Potter and Richard Birdsey

Abstract This chapter is an overview of the main site characterization requirements at landscape-scale sampling locations. The overview is organized according to multiple "Site Attribute" headings that require descriptions throughout a given study site area, leading ultimately to a sufficient overall site characterization. Guidance is provided to describe the major site attributes similarly across landscape-scale locations so that inter-site comparisons can be facilitated.

Keywords Site, characterization, climate, forest age, land use, soil

2.1 Introduction

Site characterization is required for: (1) interpreting observations and understanding processes, (2) comparing observations among study sites or with other data sets, and (3) establishing a basis for extrapolating (or scaling) from the study site to other similar sites or areas. It is desirable to characterize study sites in ways that are compatible with common standardized definitions and protocols, such as those used in national data bases. USDA's Forest Inventory and Analysis (FIA) and National Resources Inventory (NRI) are examples of long-term inventory programs that set national standards (Nusser and Goebel 1997, Bechtold and Patterson 2005). Likewise, DOE's AmeriFlux network and the National Science Foundation's (NSF) Long-Term Ecological Research (LTER) network are example of intensive

---

C. Potter
NASA Ames Laboratory, Mail Stop 242-2, Moffett Field, CA 94035
E-mail: chris.potter@nasa.gov

R. Birdsey
US Forest Service, Northern Research Station, 11 Campus Blvd., Suite 200, Newtown Square, PA 19073
E-mail: rbirdsey@fs.fed.us

C.M. Hoover (ed.) *Field Measurements for Forest Carbon Monitoring*,

monitoring programs with data requirements and standards (Law et al. 2005). While it may be impossible to completely reconcile standards and requirements among different ecosystem studies programs, an effort should be made to be as consistent as possible, even if this means classifying the same variable in more than one way.

Although there may be no universal definition of the term "site characterization" specifically for carbon cycle studies, this activity is assumed to include surveys of all critical biotic and abiotic factors of the local environment, as well as the history and impacts of human activities in the study location(s) of interest. Selected geographic and political site descriptors, such as county or congressional district, may also be useful. A "site" may be composed of more than one distinct element of a landscape, i.e., may be a mosaic of patches with distinct characteristics. Characterizing a landscape may therefore require multiple sets of site variables, each associated with a map of the site showing the distinct patches. Alternatively, some site variables may be shown as a continuum of values on a map or image of the landscape.

This chapter is organized according to multiple "Site Attribute" headings that require standardized descriptions throughout the study area, leading ultimately to a sufficient overall site characterization. Under each Site Attribute section that follows, there are lists of both the 'Basic' attribute features that should be described in a standardized manner for every landscape-scale study site, and 'Advanced' attribute features that may be collected at any study site if resources permit. Published reference and/or internet sources are included for each attribute, which provide standard data bases and methodologies for consistent site characterization results across landscape-scale locations.

2.2 General Attributes

2.2.1 Geographic and Ownership Description

Important general information about research sites includes location and ownership. Common geographic descriptors such as state and county are very useful. The study site or landscape area should be described with precise geographic coordinates (lat/long or UTM) to facilitate overlay with maps or other geographic data. The ownership of the land should be recorded using a specific agency name if the land is public, and a generic name (e.g., nonindustrial private) for private land. If the land has protected status, such as an Experimental Forest or Research Natural Area, this should be noted.

2.2.2 Present Land Use/Land Cover

Land use/land cover is a basic site classification that allows comparison or extrapolation of results to similar, broadly defined areas. Current land use/cover should be

recorded or mapped for each area using a national inventory protocol such as FIA or NRI, or a standard land cover classification such as NLCD 2001 (specify which classification schemes are used).
Recommended major land use/land cover categories and sub-categories include:
Water
Developed – residential
Developed - industrial
Forest
Cropland
Grass/field
Wetland

Advanced land-use features that indicate management intensity may include:
Agricultural management:
No-till
Fertilizer use
Irrigation
Herbicides

Forest management:
Clearcut
Partial cutting
Thinning
Site preparation
Regeneration
Fuels management
Fertilization

Hydrologic management:
Dams
Reservoirs
Canals

Suggested web sites:

FIA	http://fia.fs.fed.us/
NRI	http://www.nrcs.usda.gov/technical/NRI/
USGS Eros Data Center	http://edc.usgs.gov/products/landcover.html

2.3 Land Cover Change

Land cover changes represent one of the most significant drivers of changes in carbon stocks and fluxes. For example, a study by Casperson et al. (2001) based on analysis of forest inventory data suggested that land-use history (broadly defined to include harvesting for products and land-use change) accounted for more than 90%

of recent carbon accumulation in forests. All major historical land use changes and disturbances should be discussed in terms of their geographical distribution, frequency and intensity, the size of the area occupied by each, and the impact on land cover. The dates of the changes or disturbances should be recorded. Basic land cover change features should be characterized for all study sites, with precise dating of events if possible.

Natural disturbances:
Wildfire
Drought
Flood
Weather damage (ice, wind)
Pest outbreaks

Human disturbances:
Fire
Deforestation and/or afforestation
Timber harvest
Stand treatments (see "forest management" above)
Grazing
Agricultural treatments (see "agricultural management" above)
Introduced species

Advanced land cover change features that may be characterized include tree ring chronologies, carbon dating, and isotope characterization.

Suggested web sites:

http://landcover.usgs.gov/landcovertrends.asp

2.4 Present Vegetation Composition

Vegetation composition is associated with variability in ecosystem carbon stocks (Sun et al. 2004). Characterization of vegetation composition according to a common plant functional type is required for using site data in models, and for extrapolating results from individual sites or landscapes to larger or similar geographic areas. Major species and/or species groups are required for vegetation characterization, such that the site composition can be readily related to one of the classification systems listed below. Major plant species should be recorded in terms of geographical distribution and the size of the area occupied by each. In this section we present a few of the more common vegetation classification systems. Sometimes it is necessary to have the site classified according to more than one system, or to at least be aware of how to crosswalk from one system to another.

Forest Inventory and Analysis (FIA) – the FIA program uses a forest type classification that is relevant to land management. The system is based on a set of rules that determine forest type from the dominance or co-dominance of different tree species. There is a two-level hierarchy comprised of forest type groups and forest types. Details are available in the FIA field manual (see web site below).

Food and Agriculture Organization (FAO) – the United Nations FAO has developed a classification system that is commonly used for the world's forests, and is the basis for reporting global forest statistics in periodic Forest Resource Assessments. The FAO classification system is primarily designed to determine how "natural" forests are by determining whether trees are planted or natural, and if natural, whether the forest is "primary" (or mostly undisturbed) or "secondary" (recovering from disturbance).

National Land Cover Data (NLCD) – National land cover data is derived from Landsat satellite imagery and is classified by vegetation types that are more detailed than the basic land cover classes. Since this is a common product for mapping vegetation, it may be necessary to classify sites according to this scheme.

International Geosphere/Biosphere Program (IGBP) – MODIS and other satellite data products use the IGBP vegetation classification system, which includes 11 vegetation classes.

Advanced vegetation cover features that may be characterized include detailed descriptions of species including understory vegetation, which may also be used to construct biodiversity indices.

Suggested web sites:

FIA http://fia.fs.fed.us/
FAO http://www.fao.org/forestry/index.jsp
NLCD http://landcover.usgs.gov/index.asp
IGBP http://edcdaac.usgs.gov/main.asp

2.5 Forest Stand Age

In forest ecosystems, stand age is one of the most important variables affecting all of the carbon pools (Pregitzer and Euskirchen 2004). Stand age may not be the same as time since disturbance if the trees are of multiple ages. Time since disturbance is described in the previous section titled "land cover change". We intend *stand age* to represent the weighted mean age of live trees in the overstory. Stand age is intended to represent time since tree establishment (not time since tree reached breast height). In mixed-age stands, it may be useful to record both mean stand age and maximum tree age (age of oldest tree sampled).

Individual tree age is best measured from increment cores, although in some cases land management records may contain information about the dates of past disturbance events that may have initiated a new stand of trees. Since stand age is intended represent the mean age of live trees in the overstory, it may be estimated from as few as two to three dominant trees from the overstory of the stand (FIA 2005). These trees should be selected to represent the distribution of tree sizes and species within the overstory, and may be selected using existing information on stem diameter distribution. Investigators should use either visual inspection, or preferably, stem diameter distribution data to assign each sampled tree a weight indicating the fraction of the overstory it represents. (The sum of weights should be 100%.)

Core each representative tree at breast height (1.37 m or 4.5 ft) with an increment borer. Extract core from borer, and inspect to be sure that the core reached the tree pith. Count rings between the pith and the outside edge of the core (inside bark). Correct this age to account for years required to reach breast height, with default values of 5 years for eastern species, 5 years for western hardwoods, and 10 years for western softwoods (FIA 2003). Multiply age of sampled trees by their respective weights, and total for stand age.

2.6 Topography

Topographic information about landscapes is useful in carbon cycle studies. Soil characteristics including carbon content are strongly influenced by topographic position (Kulmatiski et al. 2004). Topography influences the movement of carbon across the landscape in soil and water. The ability to interpret the "footprint" of CO_2 flux estimates from eddy covariance measurements is strongly affected by topography (Baldocchi et al. 2000). Many ecosystem process models use elevation as a basic data layer. Topography of a region is provided by USGS topographical maps available at a variety of scales (1:5,000–1:25,000 and more). Topography may also be determined from digital elevation maps.

Basic topographic features should be characterized, including:

Ground control points (latitude/longitude/elevation) at or near the center of all measurement plot locations

List of different land forms, including plateaus, mountains, hills, and valleys

Slope classes

Advanced topographic features that may be characterized include:

Land forms of differing surface water drainage potential

Locations of streams, rivers, lakes, reservoirs

Drainage basin delineation

Relief intensity (maximum difference in elevation per sq. km, expressed in meters)

Remote imagery of the region, e.g., aerial photographs, Shuttle Radar Topographic Mission (STRM) images

Ground control points (latitude/longitude/elevation) for remote image geographic registration

Suggested web sites:

http://gos2.geodata.gov/wps/portal/gos
http://edcdaac.usgs.gov/gtopo30/gtopo30.html

2.7 Soils and Geology

In forest ecosystems, carbon in soils often represents the largest percentage of all ecosystem carbon, exceeding 75% of the total in some forest types, and averaging about 50% for U.S. forests (Heath et al. 2003). Ecosystem process models usually require information about soil characteristics. For example, a key parameter in the PnET model is soil water holding capacity (Pan et al. 2004). Thus, a good characterization of soils is an essential classification of landscape-scale carbon monitoring. Soil characterization includes a summary analysis of soil types present in the survey map products of the USDA soil taxonomy classification system. All major soils and/or categories should be discussed in terms of their geographical distribution and the size of the area occupied by each soil type, according to the USDA State Soil Geographic Database (STATSGO).
Basic soil features should be characterized for all study sites, including:
USDA soil type
Texture and soil size separates
Organic matter content
Bulk density and compaction

Advanced soil features that may be characterized include:
Texture profile distribution of soil size separates
Chemical properties, e.g., pH, CEC, base saturation
Permeability and drainage
Contamination and pollution (if relevant)

Suggested web sites

http://soils.usda.gov/
http://www.essc.psu.edu/soil_info/index.cgi?soil_data&conus

Information about parent materials and bedrock composition available in soil survey reports can be considered adequate for the purposes of site characterization. Basic geologic features should be characterized, including:
Parent material composition
Depth of soil to bedrock

2.8 Climate and Air Quality

Climate and air pollution are important drivers of processes that affect productivity and ecosystem carbon, and are almost always key input variables for a wide range of carbon models (Schaefer et al. 2002, Nemani et al. 2003, Potter et al. 2003). Climate and air quality characterization should consist of annual, seasonal, or monthly ranges and averages. Finer-scale weather characterization is covered in the chapter on micrometeorology (next chapter).

Basic climate and air quality variables should include:
Surface air temperature
Precipitation - frequency and distribution
Solar radiation flux
Relative humidity
Nitrogen deposition
Tropospheric ozone exposure

Advanced climate and air quality features that may be characterized include:
Wind direction and speed
Potential evapotranspiration
Data summaries of historical air quality in the region, in terms of ambient concentrations of hydrocarbons, carbon monoxide, nitrogen oxides, sulfur oxides, photochemical oxidants, and particulates

Suggested web sites:

http://www.ncdc.noaa.gov/oa/climate/climateresources.html

We remark in closing that the main goal for site characterization at landscape-scale sampling locations is to describe the major site attributes similarly across locations so that inter-site comparisons can be facilitated. Although investigators are also encouraged to classify site attributes in as many ways as possible as a means to more readily extrapolate across non-landscape-scale sites with varying classification systems, that is a secondary priority to well-standardized descriptions across sites in the landscape-scale network.

It is recommended that standardized worksheets (with units, wherever appropriate) be developed by investigators to include the basic and advanced attributes listed in this chapter. The worksheet templates will aid site teams in organizing their descriptions rapidly and consistently. A section for listing references should be included for each landscape-scale site description worksheet as well, which will be particularly valuable if these citations are not readily available as published papers in the literature.

Literature Cited

Baldocchi DJ, Finnigan K, Wilson KT, Paw U, Falge E (2000) On measuring net ecosystem carbon exchange over tall vegetation on complex terrain. Boundary-Layer Meteorology 96, 1–2:257–291

Bechtold WA, Patterson PC (eds.) (2005) The enhanced forest inventory and analysis program – national sampling design and estimation procedures. Gen. Tech. Rep. SRS-80. U.S. Department of Agriculture, Forest Service, Southern Research Station. 85 p.

Casperson JP, Pacala SW, Jenkins JC, Hurtt GC, Moorcroft PR, Birdsey RA (2001) Contributions of land-use history to carbon accumulation in U.S. forests. Science 290:1148–1151

Forest Inventory & Analysis (FIA) National Core Field Guide Version 3.0. (October 2005) Available at: http://fia.fs.fed.us/library/field-guides-methods-proc/docs/2006/core_ver_3-0_10_2005.pdf

Heath LS, Kimble JM, Birdsey RA, Lal R (2003) The potential of U.S. forests soils to sequester carbon. In: Kimble JM, Heath LS, Birdsey RA (eds) Carbon trends in U.S. forestlands: A context for the role of soils in forest carbon sequestration. New York: CRC Press, pp. 385–394

Kulmatiski A, Vogt DJ, Siccama TG, Tilley JP, Kolesinskasc K, Wickwire TW, Larson BC (2004) Landscape determinants of soil carbon and nitrogen storage in southern New England. Soil Science Society of America Journal 68:2014–2022.

Law BE, Baldocchi DD, Dahlman R, Davis K, Hollinger D, Munger W, Running SW, Wofsy S, Verma S (2005) AMERIFLUX strategic plan. http://public.ornl.gov/ameriflux/AmeriFlux_Strategic_Plan.pdf

Nemani RR, Keeling CD, Hashimoto H, et al. (2003) Climate-driven increases in global terrestrial net primary production from 1982 to 1999. Science 300:1560–1563

Nusser SM, Goebel JJ (1997) The National Resources Inventory: a long-term multi-resource monitoring programme. Environmental and Ecological Statistics, 4(3):181–204

Pan Y, Hom J, Birdsey R, McCullough K (2004) Impacts of rising nitrogen deposition on N exports from forests to surface waters in the Chesapeake Bay Watershed. Environmental Management 33 (4):s120–s131

Potter C, Klooster S, Myneni R, Genovese V, Tan P, Kumar V (2003) Continental scale comparisons of terrestrial carbon sinks estimated from satellite data and ecosystem modeling 1982–1998. Global and Planetary Change 39:201–213

Pregitzer KS, Euskirchen ES (2004) Carbon cycling and storage in world forests: biome patterns related to forest age. Global Change Biology 10: 2052–2077

Schaefer K, Denning AS, Suits N, Kaduk J, Baker I, Los S, Prihodko L (2002) Effect of climate interannual variability of terrestrial CO_2 fluxes. Global Biogeochemical Cycles 16(4):1102

Sun OJ, Campbell J, Law B, Wolf V (2004) Dynamics of carbon stocks in soils and detritus of different forest types in the Pacific Northwest, USA. Global Change Biology 10:1470–1481

Chapter 3
Meteorological Measurements

David Y. Hollinger

Abstract Environmental measurements are useful for detecting climatic trends, understanding how the environment influences biological processes, and as input to ecosystem models. Landscape-scale monitoring requires a suite of environmental measures for all of these purposes, including air and soil temperature, humidity, wind speed, precipitation and soil moisture, and different aspects of solar radiation. This chapter discusses sensor characteristics, including accuracy and precision, and also provides an overview of electronic data loggers, power systems, towers, and lightning protection. A list of sensors suitable for installation at landscape-scale monitoring sites and manufacturers is included.

Keywords Environmental monitoring, data logger, temperature, humidity, radiation, soil moisture, towers

3.1 Introduction

Meteorological and environmental measurements are needed at landscape-scale monitoring sites for trend detection, developing relationships between biological and climatological variables, and providing inputs to models of ecosystem biogeochemistry, land-surface exchange, or other modeling studies. Here we describe key variables to be monitored, suggest sensor accuracies and precisions, specify site requirements, and provide information about manufacturers of environmental sensors and systems. Mention of specific companies should not be construed as an endorsement by the US Forest Service. Any installation or maintenance of equipment on meteorological towers should be carried out only by qualified personnel, as should installation of such towers. In all cases appropriate safety equipment must be utilized.

D.Y. Hollinger
US Forest Service, Northern Research Station, 271 Mast Road, Durham, NH 03824
E-mail: dhollinger@fs.fed.us

C.M. Hoover (ed.) *Field Measurements for Forest Carbon Monitoring*,

Meteorological variables to be monitored at landscape-scale sites include air temperature, relative humidity, wind speed, wind direction, barometric pressure, precipitation, and direct and diffuse photosynthetically active radiation (Table 3.1). In addition, soil temperature and moisture should be quantified at several depths. If resources permit, the following additional important variables should be monitored; reflected photosynthetically active radiation, canopy wetness, and albedo. Pearcy et al. (1989) discussed general principles for making environmental measurements and describe various sensors and methods in detail. The AmeriFlux web page has additional useful information and measurement guidelines (http://public.ornl.gov/ameriflux/). See also the documentation describing international meteorological station guidelines (WMO, 1996) and the US Climate Reference Network (US Department of Commerce, 2003).

In modern meteorological stations of the type discussed here, environmental data are recorded by sensors connected to electronic data loggers. Data loggers process and store sensor data, typically generating and storing additional information such as means, maxima, standard deviations, wind vectors, etc. More advanced data loggers may be used to initiate events based on time or specific conditions such as rainfall or water level. Data loggers suited for meteorological stations are typically self-contained and capable of operating under extreme conditions. They utilize non-volatile memory so that results are not lost even if batteries fail. Data loggers typically have various communication functions so that data may be moved from the logger to a computer using direct connection, modem, or wireless capabilities. Power may come from replaceable batteries or solar panels and rechargeable batteries. Data loggers come in a range of sizes and capabilities, and are often rated by the number of sensor channels they may read. Data logger

Table 3.1 Environmental parameters to measure at landscape-scale monitoring sites with desired accuracy and precision specifications

Parameter	Accuracy	Precision	Estimated cost each ($)
I. Core parameters			
Air temperature	±0.2°C	±0.05°C	100
Relative humidity	±3%	±1%	150
Precipitation	±0.2 mm	1	500
Wind direction	5%	2°	600
Wind speed	0.5 m s^{-1}	0.2 m s^{-1}	600
Barometric pressure	±1.5 mbar	0.2 mbar	500
Solar radiation	5%	5 W	750
Total PFD	5%	2 μmol	250
Diffuse PFD	5%	2 μmol	3,000
Soil temperature	±0.2°C	±0.05°C	200 (minimum of 2)
Soil moisture	3%	0.1%	300 (minimum of 2)
II. Additional parameters			
Canopy wetness	5%	1%	100
Reflected PFD	5%	2 μmol	250
Below canopy PFD	5%	2 μmol	250
Albedo (long and short wave)	10%	1 W	5000
Data logger	0.1%	12 bit	1500

Table 3.2 Manufacturers of data loggers, environmental sensors, and tower supplies. There are several web-based resources with more comprehensive lists of instrument manufacturers. See for example http://www.meteo-technology.com/index.htm

Company	Equipment and sensors
Campbell Scientific	Data loggers, most sensors
Delta-T Devices	Data loggers, direct and diffuse radiation
Onset	Data loggers, sensors
Vaisala	Most sensors
Met-One	Temperature, humidity, wind, ventilated shields
R.M. Young	Temperature, humidity, wind
The Eppley Laboratory	Radiation sensors
Kipp and Zonen	Radiation sensors
LiCor	PFD sensors
Rohn	Towers
Nello	Towers
Isotruss	Towers
AN Wireless	Towers
Foresight Products	Earth anchors for guy wires
Northern Arizona Wind and Sun	Solar panels, deep cycle batteries, charge controllers
The Alternative Energy Store	Solar panels, deep cycle batteries, charge controllers
Mr Solar	Solar panels, deep cycle batteries, charge controllers

manufacturers are listed in Table 3.2. For the sensor configuration suggested here, data loggers with 12 or more channels are necessary. For meteorological data, sensors need be sampled no more than once every few seconds. We recommend that individual readings from most sensors (except precipitation) be averaged over a 30-min time period for final use and that the data logger also be configured to record the standard deviation of the half-hourly readings. Precipitation data should be summed over the 30-min time interval.

3.2 Sensor Selection

Sensors are quantified via their accuracy, precision, and stability (Table 3.1). Accuracy refers to how close the reading from a sensor may come to the "true" value while precision indicates the smallest change that may be accurately recorded. The output from a sensor may degrade or "drift" over time or the sensor may respond in part to an environmental factor other than that to which it was designed to respond (typically temperature affects the response of most sensors). The stability of a sensor is a quantification of the potential for the sensor output to vary with time or other (non-measured) factor. Generally, sensor accuracy, precision, and stability all increase with price. Because long-term stability and high accuracy are not requirements of overriding concern for landscape-scale monitoring, standard meteorological sensors available from any of the manufacturers listed in Table 3.2, as long as they meet the accuracy and precision guidelines in Table 3.1, are suitable. Table 3.3 lists examples of sensors that are suitable for landscape-scale monitoring.

Table 3.3 A prototype installation for a forest landscape-scale monitoring site. Note that sensors from other manufacturers meeting the specifications in Table 3.1 are suitable

Sensor	Manufacturer and model	Location
I. Basic installation		
Air temperature and RH	Vaisala model HMP45C	5 m above canopy
Precipitation	Texas Electronics model TR-525	Above canopy
	or Vaisala model QMR102	In clearing
Wind direction and speed	Met One model 014A/024A or RM Young model 05103	On boom, 5 m above canopy
Barometric pressure	Vaisala model PTB210	With data logger
Total and diffuse PFD	Delta-T model BF3	Tower top (avoid shading)
Solar radiation[a]	Hukseflux model LP02 or Kipp and Zonen CMP3	Tower top (avoid shading)
Soil temperature	Campbell model 107	5 and 50 cm depth
Soil moisture	Campbell model CS616	5 and 50 cm depth
Data logger	Campbell model CR1000	In enclosure accessible from ground
II. Additional sensors		
Canopy wetness	Campbell model 237	Mid canopy
Reflected PFD	LiCor model LI190SB or Kipp and Zonen model PAR lite	Requires up and down facing above-canopy sensors on 2 m boom towards equator
Below canopy PFD	Kipp and Zonen model PAR lite or LiCor model LI190SB	
Albedo[a] (4 component)	Hukseflux model nr01 or Kipp and Zonen CNR1	On 2 m boom towards equator
Data logger	Campbell model CR3000	In enclosure accessible from ground

[a]Solar radiation sensor not needed when albedo measured

Because all sensors degrade in time and with exposure to the weather, it is important to calibrate or replace sensors on a regular basis following the manufacturer's recommendations. The easiest way to do this is to substitute a spare instrument and send the old one in for calibration at the prescribed interval. Purchasing spare sensors for this purpose should be part of every meteorological equipment budget. Calibration procedures and standards employed should be documented, as should the calibration results of each sensor.

3.2.1 Temperature and Humidity

Biological and physical climate processes are generally sensitive to temperature, and it is a required input to virtually all land surface and biogeochemistry-based models. Temperature sensors rely on different principles but as long as they meet the desired requirements of accuracy and precision (Table 3.1) the particular principle employed is of no importance. Air temperature sensors, however, are susceptible to heating by direct solar radiation, causing errors in the measurement. The recommended solution is to locate the sensor in a ventilated shield where the ventilation is accomplished by a fan. For remote systems, a 12-V DC fan may be powered directly off a small (~5 W) solar panel for this purpose (see later section on power systems). Temperature sensors installed into the soil should be electrically insulated to prevent the development of "ground loops", spurious currents that affect sensor readings. Use of differential inputs and good grounding practices will also help eliminate ground loops (usually data logger manuals have specific information about these topics).

Humidity combined with temperature is an important determinant of evaporation rate. Plant physiological processes such as stomatal conductance are also sensitive to humidity. Generally the simplest way to measure humidity is to use a combination temperature and humidity sensor. For accurate readings these sensors should always be used with a ventilated shield.

3.2.2 Precipitation

Precipitation data are used in conjunction with other data in many models to calculate water balance and drought parameters. "Tipping bucket" type rain gauges are best suited for long-term unattended monitoring. In these gauges water is collected by an integrated funnel and channeled to a small mechanism that "tips" and sends a pulse to the data logger for each increment of rain (such as each 0.2 mm). However, several problems must be recognized with these sorts of remote systems. First, in many climates, low temperatures cause winter precipitation to fall as snow or sleet and simple tipping bucket type rain gauges will fail to detect these events, instead falsely recording precipitation when temperatures rise sufficiently for trapped precipitation to melt. Other alternatives (heated or antifreeze-type weighing gauges) are generally more expensive and difficult to service. A heated gauge also usually requires AC line power although gauges that use propane for heating are available. However, newer load cell based weighing gauges that do not require heating (e.g. Vaisala model VRG101) show promise. A second problem with tower rain gauge installation is wind shielding. It is always recommended that a wind shield be installed around the orifice of a rain gauge to reduce errors associated with wind-driven precipitation but this is generally impossible in a tower-mounted configuration. For these reasons precipitation data from tipping bucket type rain gauges, especially in the winter, must be considered less reliable than ordinary meteorological service data.

3.2.3 *Radiation*

Solar radiation provides the energy for photosynthesis and drives the climate system through its evaporation of water and heating of the air. Solar radiation is a critical input for virtually all models of biogeochemistry or land surface fluxes. Solar radiation consists of both incident and reflected components and the incident radiation may be further divided into direct (emanating from the solar disk) or diffuse (scattered from the sky and clouds) radiation. Another way to divide solar radiation is by wavelength, typically consisting of shortwave (typically 0.285–2.8 μm in wavelength) and longwave infrared radiation (2.8– ~ 50 μm wavelength). Shortwave radiation consists of UV (0.285–0.4 μm), photosynthetically active radiation (0.4–0.7 μm), and near infrared radiation (0.7–2.8 μm). Net radiometers measure the difference between the incident and reflected short- and long wave radiation. The most useful type measures all four components (incident and reflected short- and longwave radiation) individually allowing calculation of the surface albedo, but these instruments are expensive (~$5,000). Such data are very useful for energy balance studies, and quantify solar and longwave energy fluxes in units of watts (W) m^{-2}.

Photosynthesis is most directly related to the quantity of photons intercepted. This measure of solar radiation is termed photosynthetically active photon flux density (PFD), and is measured in micromole photons per square meters per second between 0.4 and 0.7 μm. Diffuse radiation is used more efficiently by canopies than direct radiation, so the separation of total PFD into direct and diffuse components is valuable. Another useful measure is to record data from PFD sensors installed below as well as above a canopy thus allowing the fraction of incident radiation absorbed by the canopy to be estimated. Because of spatial heterogeneity in plant canopies it is usually desirable to employ many below-canopy sensors when these data are desired.

3.2.4 *Wind Speed and Direction*

These data are generally used in land surface models and for climatological studies. Wind speed and direction have been traditionally measured by rotating cup and vane systems but require threshold speeds to be exceeded before they provide reliable readings. They are also susceptible to environmental damage from hail and ice. New stationary wind speed and direction probes that measure the transit time of ultrasonic sound pulses are available, and these are generally more reliable, and cost competitive, with older designs.

3.2.5 *Soil Moisture*

Plant physiological processes and microbial activity are sensitive to the effects of too much or too little soil moisture. This can be a difficult and somewhat expensive measurement, as soil moisture should be measured at several depths. We recommend at least near surface (5 or 10 cm) and mid-profile (50 cm) depths, and duplication of sensors in several profiles is desirable. Many soil moisture probes include integral temperature sensors. Older style systems based on electrical conductivity (e.g. gypsum blocks) should not be used because of calibration problems. Newer sensors rely on different principles such as the soil dielectric constant that is more directly related to total soil water content.

3.3 Power Systems

The power requirements for data loggers and meteorological sensors are generally modest. Most loggers will operate for many months on one set of replaceable batteries. However, solar panels and rechargeable batteries provide greater reliability and flexibility. If higher current devices are in use such as fans for ventilated shields or heaters for radiometers or rain gauges, the best practice is to install a second solar power system that is independent of the data logger. In this way a power shortage resulting from a series of overcast days will only affect the high current devices.

For a simple logger setup, a small solar panel (5–10 W) connected to a small (6–12 Ah capacity) rechargeable battery should provide a reliable supply. Many loggers such as those from Campbell Scientific have optional rechargeable batteries and charge controllers built in. We recommend an additional 5-W solar panel be connected to a 12 V DC fan in the ventilated temperature shield. No battery is necessary for this installation because the problem of solar heating of the temperature sensor only occurs when the sun is out.

For greater current demands more solar panel and battery capacity are necessary, with the specific capacities depending upon climate and desired reliability. For modest demands (0.5 A at 12 V) with good reliability (5 cloudy days operation) in the northern US, a system consisting of 80 W of solar panels, 15 A charge controller, and 120 Ah of deep cycle storage batteries would be adequate. Pulse type solar controllers may cause interference with data loggers so complete separation of the logger and high power charging systems (including separate grounds) is necessary.

To optimize power when it is least available (midwinter) panels need to be inclined steeply. The general rule is latitude plus 23.5°.

3.4 Lightning Protection

In many parts of the United States lightning can be an important hazard to field meteorological sites. The data loggers available from the manufacturers listed in Table 3.2 generally include built-in lightning protection (from spark gaps or other devices) that works well *if* the datalogger and tower are grounded properly (follow manufacturer's directions). One should always avoid running wires along the ground for long distances (e.g. >30 m) because high voltages may be induced into such lines even in the absence of nearby lightning strikes. Additional protection of sensor and data lines using gas discharge tubes, tranzorbs, and/or lightning fuses is always recommended.

3.5 Sensor Installation

The World Meteorological Organization (WMO 1996) provides guidance for the installation of meteorological sensors at conventional climatological sites. The recommended sensor height (1.5 m above the ground except higher at sites which typically experience significant snow) is consistent with the recommended height for sensors in the US Climate Research Network, the Automated Surface Observing System (ASOS), Automated Weather Observing System (AWOS), and US Cooperative Observing Network (COOP). If the landscape monitoring site is dominated by short-stature vegetation or a mostly open canopy (projected canopy coverage is less than one-third of the ground area) then the 1.5 m sensor height should be used for temperature, humidity, precipitation, and wind speed data. For open sites that experience significant snow cover, sensors are to be installed at a height of 0.6 m above the surface of the average maximum snow depth or 1.5 m above the surface of the ground (no snow), whichever is higher. Radiation sensors should be mounted at the top of the sensor support structure or on a boom that extends toward the equator to avoid shadows on the sensors.

There are no standard guidelines for sensor installation at sites with partial or complete plant canopy coverage. We recommend in this instance that meteorological sensors be located 5 m above the average canopy height, which may require the installation of a meteorological mast or tower.

3.6 Meteorological Towers

Towers suitable for micrometeorological investigation are available from several manufacturers (Table 3.2). These are generally modular in design so that each section attaches to the one below. Towers are rated by their wind and ice load capabilities and because of the unpredictable nature of a forest environment (branch or tree fall upon a tower or guy wire), we recommend purchasing towers with the highest available wind load rating (typically 120 mph). Towers must be installed by qualified personnel and according to the manufacturer's instructions. Requirements for grounding should be strictly followed. Towers should never be erected near overhead electrical lines. Towers should only be climbed by trained personnel utilizing appropriate, OSHA-approved, safety equipment. Tower construction and safety courses are available from e.g. ComTrain, LLC.

Literature Cited

Pearcy RW, Ehleringer J, Mooney HA, Rundel PW (eds.) (1989) Plant Physiological Ecology. Field Methods and Instrumentation, Chapman & Hall, London

US Department of Commerce (2003). United States Climate Reference Network (USCRN) Functional Requirements Document. NOAA/NESDIS CRN Series X040. NOAA-CRN/OSD-2003-0009R0UD0 June 27, 2003 DCN 0

WMO (1996) Guide to Instruments and Methods of Observation, Geneva, Switzerland, Doc 8

Part II
Measuring Aboveground Carbon Pools

Chapter 4
Estimating Aboveground Carbon in Live and Standing Dead Trees

Peter S. Curtis

Abstract Carbon contained in aboveground live and standing dead tree mass can represent less than 1% to over 60% of the total forest ecosystem carbon pool. Accurate assessment of this pool is important for many forest managers and ecologists. Aboveground tree mass is estimated using published allometric equations relating diameter at breast height to total tree mass or component parts of bole wood, bark, branch, and foliage mass. Care must be taken in selecting appropriate allometric equations, considering the range of tree sizes to be measured, and applying logarithmic regression correction factors. Substantial error in tree carbon mass estimation can result if allometric equations are used inappropriately. Standing dead tree mass is estimated using live tree allometric relationships if the tree is recently dead and intact or from stem volume estimation and measurements of wood density if the crown is broken. Field measurement guidelines, sample calculations, and special considerations for irregular stem configurations are presented.

Keywords Allometry, logarithmic regression, snag

4.1 Introduction

The carbon contained in the aboveground parts of trees often constitutes a substantial fraction of the total carbon (C) pool in forested ecosystems, in some temperate and tropical forests equaling or exceeding belowground C pools (Pregitzer and Euskirchen 2004). However, climate, species composition, soil fertility, and the history of natural or anthropogenic disturbance all can play major roles in determining

P.S. Curtis
Department of Evolution, Ecology and Organismal Biology, The Ohio State University, 318 W. 12th Avenue, Columbus, OH 43210
E-mail: curtis.7@osu.edu

the relative importance of aboveground tree tissues in the ecosystem carbon budget. Forested boreal peatlands, for example, may have less than 1% of total ecosystem C in aboveground living tree mass (Bhatti et al. 2006). At the opposite end of the spectrum, Homann et al. (2005) reported that more than 60% of total ecosystem C was contained in the aboveground tree fraction of old-growth coniferous forests in the central Cascades of North America. In a late-successional, aspen-dominated forest typical of much of the upper Great Lakes region of North America, Gough et al. (2008) found 41% of ecosystem C in aboveground tree boles and foliage (73.5 Mg C ha^{-1}), about 9% less than was present in soil organic matter.

Accurately estimating the volume or mass of merchantable wood has been an objective of forest mensuration science for many decades (Husch et al. 2003) and a large body of basic and applied information on its application is available for many types of forested ecosystems (e.g., FAO 1989). As the goals of land managers and ecologists have expanded to include the study and management of whole-system C stores, obtaining an accurate measure of the size of this key aboveground C pool has not diminished in importance. Fortunately, the measurement and analytical methods established for calculating timber harvest yields are readily transferable to the needs of the forest C assessment community. However, the range of application of these methods is likely to be considerably broader than ordinarily encountered in commercial forestry, encompassing ecosystems of greater species diversity, tree age and size distribution. It is important, therefore, that those intending to estimate aboveground C in live and standing dead trees be fully aware of the strengths and limitations of these approaches.

Estimating the total aboveground dry mass of an individual live tree i of species $x\left(\hat{M}_{xi}\right)$ is done with the use of allometric equations of the form $\hat{M}_{xi} = aD^{b}_{xi}$, where D_{xi} is diameter at breast height (1.37 m) of that individual and a and b are parameters obtained from curves fitted to prior harvest data. D_{xi} also may be related allometrically to the mass of separate aboveground components such as bark, bole wood, branches, and leaves. Plot-level aboveground mass (Mp) is calculated by summing over n individuals of m species measured in that plot: $M_p = \sum_{i=1}^{n}\sum_{x=1}^{m}\hat{M}_{xi}$. The aboveground dry mass of standing dead trees, or snags (*sensu* Harmon and Sexton 1996) also may be estimated using allometric equations if the tree is recently dead and has not broken up. If that is not the case, snag mass is estimated from measurements of wood volume and density in a manner analogous to estimating the mass of individual pieces of downed woody debris.

4.2 Allometric Equations

There are two general points of concern in using allometric equations to estimate tree mass. The first regards the equations themselves while the second involves the field measurements.

There are several important points to be considered in the selection and use of allometric equations. The first is purely mathematical. For computational reasons, the parameters a and b often are calculated from linear regression of the log transformed independent (D) and dependent (M) variables obtained from destructive harvests of trees from a range of diameters: $\ln M = \ln a + b \ln D$. Application of this equation to predict the logarithmic mass of an individual i $(\ln M_i)$ is straightforward: $\ln \hat{M}_i = \ln a + b \ln D_i$. However, because of the nature of logarithmic regressions, the anti-log of $\ln \hat{M}_i$ is not an unbiased estimate of the arithmetic value $\hat{M}_i$. That is, $\hat{M}_i \neq e^{\ln \hat{M}_i}$. Baskerville (1972) suggested a correction factor (C) for this bias equal to the anti-log of one-half the sample variance (s^2). That is, $\hat{M}_i = e^{\ln \hat{M}_i + 0.5s^2}$, or $\hat{M}_i = C * e^{\ln \hat{M}_i}$. Correction factors should be published along with the allometric parameters a and b if these were calculated from logarithmic regressions. Note, that if a and b were estimated from non-linear regression against untransformed data, a correction factor is not needed.

A key consideration is selection of the actual allometric equations to be used. Since both species identity and site characteristics affect allometric relationships, the more species- and site-specific the equations, the better. There now are published allometric equations for many North American tree species as well as more generalized equations for larger taxonomic divisions or functional types (e.g., Perala and Alban 1993, Ter-Mikaelian and Korzukhin 1997, Jenkins et al. 2003). In many cases published allometric equations allow the separate estimation of the mass of bark, bole wood, branches, and leaves. Although the cost of developing new allometric equations is high (see Gower et al. 1997 for a description of methods) and generally not expected as part of landscape-scale monitoring activities, the potential error in estimating stand mass from the use of inappropriate allometry is large. For example, Clark et al. (2001) estimate errors of from −20% to +11% on calculated mass of tropical trees through the use of allometry developed off-site. Therefore, if equations are used that were derived from different species or from trees growing under very different site conditions, it is highly desirable to conduct limited destructive harvests to test their accuracy.

A corollary to this guideline is that attention must be given to the range of diameters used in developing the allometric equations. Because of the effort involved in harvesting large trees, these are often underrepresented in allometry datasets. Conversely, if allometric equations were developed with a focus on merchantable wood, small diameter trees may not be included. If the diameters of trees in sample plots fall appreciably above or below those harvested for the allometric determinations, large errors in $\hat{M}_i$ are possible. If this is the case, limited harvests to supplement existing datasets may be advised.

4.3 Estimating Snag Mass

A snag is defined as a standing dead tree with an angle of lean of 45° or less from the vertical. Those leaning more than 45° are considered part of the downed woody debris pool and are treated in Chapter 6. If the snag is not broken one can estimate its mass from measurements of D and the use of appropriate allometric equations as described above for live trees. If the crown is broken one must estimate the remaining bole wood volume (V) by measuring the snag height (H) and area at its base (A_b) and top (A_t). Height can be measured on tall snags using a clinometer and tape or on smaller snags using a meter stick. The A_t is measured from remains of broken material found on the ground or estimated visually. Two different volume equations are appropriate from these measurements depending on whether the snag volume more closely approximates the frustum of a cone, where:

$$V = H * (A_b + (A_b * A_t)^{0.5} + A_t) / 3 \tag{4.1}$$

or if it approximates the frustum of a paraboloid, where:

$$V = H * (A_b + A_t) / 2 \tag{4.2}$$

Converting V to $\hat{M}_i$ requires knowledge of the density of the snag wood which is in turn a function of snag species and decay class. Density data is available from the literature for some species (e.g., Harmon and Sexton 1996) but may need to be determined on site from tree cores.

4.4 Mass to Carbon Conversion

The C content of woody mass may vary across constituent tissues such as bark, sapwood, and heartwood. For example, in *Betula papyrifera*, outer bark C content is ~68% compared to ~49% in sapwood and heartwood. However, outer bark constitutes <5% of total aboveground woody mass in mature individuals of this species, so that whole wood C content may be satisfactorily approximated by that of the sapwood and heartwood, i.e., 49%. Where bark constitutes a large faction of aboveground woody mass it may be necessary to calculate a mass-weighted average whole wood C content. This also requires the use of allometric equations that partition bark from bole wood mass. If published estimates of wood C content for a given species are not available it is recommended that elemental analysis of bark, sapwood and heartwood be conducted from tree cores taken at breast height. Dead wood C content by decay class also is needed to convert snag wood mass to snag C mass. We found little change in wood C content across decay classes in *Populus* and *Betula* but this should be confirmed on a species-specific basis.

4.5 Field Measurements

Landscape-scale monitoring methods for measuring tree diameter at breast height (D) follow those of the USDA Forest Inventory and Analysis Program (2002). In addition, to allow accurate remeasurement, all trees within a plot should be tagged with numbers enabling unique identification of all individuals > 10 cm D. Use standard aluminum forestry labels with tree ID information entered using a character punch and fixed to the bole with an aluminum nail. To guard against loss of information if tags disappear, all individuals > 10 cm *D* within a plot should also be spatially located by measuring the distance and compass bearing to a permanent marker with known GPS coordinates.

Diameter measurements are made 1.37 m above the ground on the uphill side of the tree using diameter tape ruled in metric units. To facilitate remeasurement, the point of measurement should be marked with a scribe, crayon, paint, or aluminum nail (note, however, that a nail may cause splitting and callus tissue formation in some species rendering subsequent measurements inaccurate). Special considerations arise under the following circumstances: forked trees, stump sprouts, trees with surface irregularities, leaning trees, live windthrown trees, and downed trees with tree-form branches growing from the main bole. In addition, for some woodland species such as mesquite and juniper, mass allometry is based on diameter at root collar. General guidelines only are given below. For a detailed discussion and illustration of these special considerations, see Volume 1 of the Forest Inventory and Analysis National Core Field Guide (U.S. Department of Agriculture, Forest Service 2005).

Trees forked below 1.37 m are treated as separate trees, those forked above 1.37 m are treated as a single tree. Forks originate at the point on the bole where the piths intersect. Stump sprouts are handled the same as forked trees. For trees with irregularities at 1.37 m such as swellings, bumps, depressions, and branches, measurements are made immediately above the irregularity. Leaning trees are measured 1.37 m along the underside face of the bole, while live windthrown trees are measured from the top of the root collar. For downed trees with vertical, tree-form branches, if the pith of the main bole is above the duff layer, the forking rules for forked trees apply. If the pith is below the duff layer, each tree-form branch is treated as a separate tree. Diameter at root collar is measured below the litter layer but above the mineral soil, and above any swells present. For multi-stemmed woodland trees, diameter at root collar is computed as the square root of the sum of the squared stem diameters.

4.6 Examples

Field records for live tree data are straightforward, generally including only date and plot ID, individual tree ID (inscribed on permanent tag), species code, distance and angle from a fixed marker, and *D*.

Calculations for estimating the C content of the aboveground woody mass of an individual of *Fagus grandifolia* where D = 40.5 cm are shown below. Using the allometric equation $\hat{M}_{xi} = aD_{xi}^{b}$, a = 0.1892 and b = 2.3097 (Young et al. 1980, range 3–66 cm):

$$\hat{M}_{xi} = 0.1892 * 40.5^{2.3097} = 976 \text{ kg woody dry mass}$$

$$976 \text{ kg dry mass} * 0.49 \text{ kg C/kg dry mass} = 478 \text{ kg C}$$

Literature Cited

Baskerville GL (1972) Use of logarithmic regression in the estimation of plant biomass. Journal of Forest Research 2: 49–53

Bhatti JS, Errington RC, Bauer IE, Hurdle FA (2006) Carbon stock trends along forested peatland margins in central Saskatchewan. Canadian Journal of Soil Science 86: 321–333

Clark DA, Brown S, Kicklighter DW, Chambers JQ, Thomlinson JR, Ni J (2001) Measuring net primary productivity in forests: concepts and field methods. Ecological Applications 11: 356–370

FAO (1989) Studies on the volume and yield of tropical forest stands. Food and Agriculture Organization of the United Nations, Rome.

Gough CM, Vogel CS, Schmid HP, Curtis PS (2008) Controls on annual forest carbon storage: Lessons from the past and predictions for the future. Bioscience 58: 609–622

Gower ST, Vogel JG, Norman JM, Kucharik CJ, Steele SJ, Stow TK (1997) Carbon distribution and aboveground net primary production in aspen, jack pine, and black spruce stands in Saskatchewan and Manitoba, Canada. Journal of Geophysical Research 102: 29,029–29,041

Harmon ME, Sexton J (1996) Guidelines for Measurements of Woody Detritus in Forest Ecosystems (US LTER Publication No. 20). US LTER Network Office, University of Washington, Seattle, WA

Homann PS, Harmon M, Remillard S, Smithwick EAH (2005) What the soil reveals: Potential total ecosystem C stores of the Pacific Northwest region, USA. Forest Ecology and Management 220: 270–283

Husch B, Beers TW, Kershaw JA Jr. (2003) Forest Mensuration. Wiley, Hoboken, NJ

Jenkins JC, Chojnacky DC, Heath LS, Birdsey RA (2003) National-scale biomass estimation for United States tree species. Forest Science 49: 12–35

Perala DA, Alban DH (1993) Allometric biomass estimators for aspen-dominated ecosystems in the Upper Great Lakes. Research Paper NC-314. St. Paul, MN: U.S. Dept. of Agriculture, Forest Service, North Central Forest Experiment Station, 38 p.

Pregitzer KS, Euskirchen ES (2004) Carbon cycling and storage in world forests: biome patterns related to forest age. Global Change Biology 10: 2052–2077

Ter-Mikaelian MT, Korzukhin MD (1997) Biomass equations for sixty-five North American tree species. Forest Ecology and Management 97: 1–24

U.S. Department of Agriculture, Forest Service (2005) Forest inventory and analysis national core field guide, volume 1: field data collection procedures for phase 2 plots, version 3.0. U.S. Department of Agriculture, Forest Service, Washington Office. Internal report. On file with: U.S. Department of Agriculture, Forest Service, Forest Inventory and Analysis, Rosslyn Plaza, 1620 North Kent Street, Arlington, VA 22209

Young HE, Ribe JH, Wainwright K (1980) Weight tables for tree and shrub species in Maine. Life Sciences and Agriculture Experiment Station, University of Maine at Orono. Miscellaneous Report 230, 84 pp.

Chapter 5
Measuring Carbon in Shrubs

David C. Chojnacky and Mikaila Milton

Abstract Although shrubs are a small component of the overall carbon budget, shrub lands and shrub cover within forested lands warrant monitoring with consistent procedures to account for carbon in shrubs and to track carbon accumulation as communities change from shrubs to trees and vice versa. Many different procedures have been used to sample and measure shrubs (Bonham 1989) but only three types are selected here, to represent a range from simple and subjective to more time-consuming but objective measurements. Although the goal is to measure shrub carbon, the methods outlined here estimate biomass—which is about 50% carbon. For sample design, we advocate compatibility with the USDA, Forest Service, Forest Inventory and Analysis (FIA) program by using transects, microplots, or quadrats arranged within or near FIA subplots. Three basic methods are suggested for measuring shrub biomass: (1) cover estimations along transects, including point-intercept and line-intercept; (2) visual cover estimates in fixed area units; and (3) diameter measurement within fixed-area sampling frames. The 3rd method for measurement of individual shrub stem-diameters provides the most robust data for estimating biomass (and by extension, carbon) but requires the most field time. The other two methods allow more rapid measurements of shrub cover along transects or within plots. Our summary provides a framework for collecting shrub measurements three different ways; however, more work will likely be needed to develop appropriate equations that equate cover or stem measurements with biomass for various species.

Keywords Cover measurement, diameter measurement, microplot, shrub, transect sampling

D.C. Chojnacky
US Forest Service Washington, DC, Current address: Virginia Tech Department of Forestry, 144 Rees Place, Falls Church, VA 22046
E-mail: dchoj@cox.net

M. Milton
National Science Foundation, Arlington, VA, Current address:
National Park Service, 1900 Anacostia Drive SE, Washington, DC 20020
E-mail: mikaila_milton@nps.gov

C.M. Hoover (ed.) *Field Measurements for Forest Carbon Monitoring*,

5.1 Introduction

The movement of carbon in its many forms – between the living and nonliving forms of the biosphere, atmosphere, oceans, and geosphere – is widely recognizable as the carbon cycle. With the amount of carbon in the atmosphere escalating at a rapid rate today, management focus is increasingly turning to the sequestration of more carbon where possible. In forests, shrubs are a small component of the overall carbon budget, estimated as 2% of total forest carbon (Kimble et al. 2002). However, shrubs often dominate early successional stages of many forest types, particularly following fires, and in some cases vigorous shrub communities constitute a primary land management objective for wildlife cover and forage. Furthermore, shrub and other understory growth (net primary production) can be comparable to that of trees (Nilsson and Wardle 2005) and constitutes a major source of carbon for the forest floor and soil. Therefore, shrubs may not sequester much carbon on land but they may be very important for adding to the soil carbon pool.

Consequently, from a carbon stock perspective, shrub lands and shrub cover within forested lands warrant monitoring with consistent procedures to account for carbon in shrubs and to track carbon accumulation as communities change from shrubs to trees and vice versa.

Procedures to measure forest vegetation differ depending on the life forms of live and dead vegetation. Typically, either fixed-area plots or transects are used to sample vegetation material; estimates generally are expressed in tons or Mg per unit area. Shrubs have been successfully sampled with both plots and transects (Peet et al. 1998, Korb et al. 2003, McMillin and Allen 2003).

Sampling method definitions
Transect ~ line used to survey the distribution of organisms across a given area, using sample points along the line to collect data that can be used to describe the whole population area.
Fixed-area plot ~ plot (circular, square, or rectangular) of a fixed size used to survey a portion of a forest or other area to collect data that can be used to describe the whole population area.

Elsewhere in this book, procedures developed by the USDA, Forest Service, Forest Inventory and Analysis (FIA) program (FIA 2005) are recommended for measuring carbon in forest trees. However, the FIA shrub measurements were not designed for measuring carbon. Therefore, in this chapter we recommend shrub measurement procedures that are compatible as an add-on to the FIA protocol and are more suited to the measurement of shrub carbon.

Many different procedures have been used to sample and measure shrubs (Bonham 1989) but only three types are selected here, to represent a range from simple and subjective to more time-consuming but objective measurements. These methods incorporate two different sample-design philosophies: transects and fixed-area

Defining shrubs

A shrub is a woody plant that generally has multiple basal stems growing from the same root system. The woody growth form distinguishes a shrub from non-woody herbaceous plants. However, it is sometimes difficult to distinguish shrubs from trees based on growth form, so it is usually necessary to consult formal species lists, such as the comprehensive list of U.S. trees compiled by the FIA (FIA 2004a). The FIA list is a good starting point for deciding whether a woody plant is a tree or shrub for each project. Nevertheless, a few woody species that FIA calls a "tree" might be better measured as a "shrub" such as some species of *Amelanchier*, *Prunus*, and *Salix*. Therefore, shrub definitions for individual projects should be established at the outset of any monitoring program. In addition to choosing which species are defined as shrubs, it will be practical to set a lower basal diameter limit below which individuals are considered to be "herbaceous" for measurement purposes. These small-diameter plants are often more efficiently measured with the herbaceous layer (which is not covered in this chapter).

plots. Generalized equipment lists, procedures, calculations, and discussion are provided for comparison purposes. References are provided for details on implementation of various methods.

Although the objective is to measure shrub carbon, none of the methods outlined here estimates carbon directly. Instead, either shrub cover or basal area is measured in the field, and biomass—which is about 50% carbon (Kort and Turnock 2003)—is estimated from regression subsampling (Lohr 1999) or from auxiliary models. We provide an overview of some existing auxiliary models for estimating shrub biomass and show how the model equations were applied during a 2004 test within piedmont hardwood forests in the eastern United States.

5.2 Incorporating Shrub Measurements into the FIA Design

FIA procedures have become the standard for strategic-scale sampling of forests in the United States. FIA has more than 70 years of experience monitoring wildland and managed forests and more than 10 years of experience monitoring forest health indicators, which include shrub cover measurements. The FIA program already covers all forest lands in the US with a grid of 120,000 plots and is currently adapting its forestland procedures for use in urban forest health monitoring efforts.

The FIA design includes fixed-plot procedures for measurement of understory vegetation diversity and structure, including shrub species (FIA 2002). Shrubs are measured as cover in a nested design as part of the phase 3 (P3) or Forest Health Monitoring (FHM) protocol. In FHM plots, cover is visually estimated over each subplot as well as in three 1 m^2 quadrats within each subplot. In the subplots, cover is recorded in classes (1–5%, 6–10%, 11–20%, 21–40%, 41–60%, 61–80%, and 81–100%) for all species including shrubs. In the quadrats, cover is estimated to the nearest 1% for each species.

Within each FIA subplot, a 2.1 m radius microplot is used to sample seedlings and saplings and is used by some regional FIA units for various understory measurements (Fig. 5.1) The FIA design also includes transect sampling procedures for measuring down deadwood on P3 FHM plots. Three transects radiate out from each subplot center at 120° angles out to the 7.3 m distance.

We advocate the use of microplots, quadrats, and transects arranged within or near FIA subplots to link the shrub carbon measurements to current FIA design. Shrub measurements do not need to be taken on existing microplots or transects (to avoid possible over trampling) but could be accomplished by adding more microplots or transects that could be offset from deadwood transects (by 60°, for example).

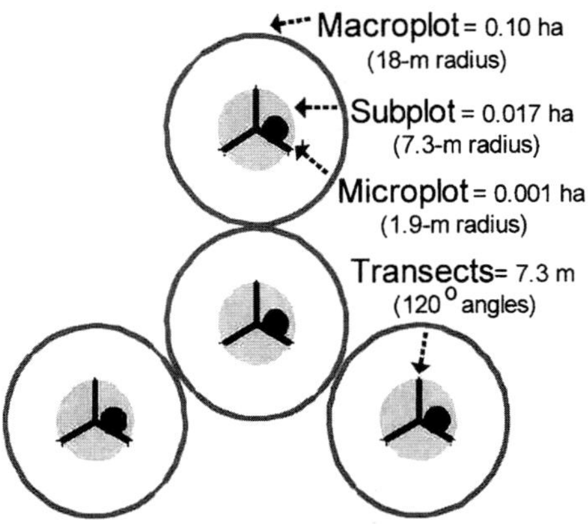

Fig. 5.1 The FIA plot design includes a cluster design of four nested plots. Most tree measurements are made in the 0.017-ha subplots; however, seedlings, saplings, and understory are measured within the 0.001 ha microplots or within 1-m square quadrats placed along transects in the subplots. Deadwood is measured on transects. See Bechtold and Patterson (2005) and FIA (2004b) for details

Although it is common to sample vegetation composition and structure at different scales (Peet et al. 1998), such methods were designed to capture community structure and to track rare and alien species rather than to provide an objective measure for shrub carbon. Furthermore, methods using variable-size sampling frames for different vegetation structures have not been widely adapted for biomass or carbon estimation.

Therefore, each shrub measurement method described in this chapter uses a single-size sample frame for each plot. However, the various methods presented still offer flexibility for measuring different plant communities. Our assumption is that the primary objective is to achieve an unbiased estimate of total shrub carbon, rather than to monitor community composition over time or to sample all plant species for the purpose of detecting rare plants or invasive species.

5.3 Shrub Sampling Methods

We focus on field estimation of shrub cover (or a cover surrogate) and on measurement of basal diameters (basal area) to provide the data for estimating biomass through the use of regression subsampling (Lohr 1999, p. 74) or auxiliary models. Carbon in shrubs is assumed to constitute about 50% of biomass (Kort and Turnock 2003). Cover is defined as the vertical projection of vegetation from the ground as viewed from above (Elzinga et al. 1998). Shrub measurements are presented in the context of sampling points along a transect or fixed-area sampling units, as part of a larger survey where plots constitute sample units that are aggregated in classical sample designs for estimating population attributes (Cochran 1977). The objective of these shrub sampling techniques is to adequately sample a plot for shrub cover or basal area. In terms of overall sample design, these plots can be viewed as measuring cover or basal area in a "support region" around a point, which is designed to be large enough to capture the variation at that point and represent a true estimate of the parameters measured (Chojnacky 1998, p. 3). The shrub data collected as part of the plot measurement protocol can then be combined with other carbon components sampled around the same point without necessarily using exactly the same area dimensions for each element of the vegetation.

Plot dimensions in typical vegetation studies (such as those conducted by FIA) have several sample size features that are critical to the performance of each of sample method; those features include transect length, plot or quadrat size, and number of sample measurements. However, no attempt is made in this chapter to establish optimum dimensions for each feature; rather, initial recommendations are given based on a field test conducted in eastern deciduous forests in 2004 (described later in this chapter). To fully optimize application of the procedures presented here, additional field testing is suggested for each major ecosystem or area of the US.

Three basic methods are suggested for measuring shrub biomass: (1) cover estimations along transects, including point-intercept and line-intercept; (2) visual cover estimates in fixed area units; and (3) diameter measurement within fixed-area sampling frames. Choice of method to use depends on a project's objectives, available staff time, funds available, level of precision and degree of confidence desired, type of vegetation being sampled, season, and other site-specific factors. Regardless of method, it is best to sample during the period of peak foliage development to ensure consistency among samples and to avoid complicated adjustments that otherwise would be needed to account for seasonal leaf conditions.

5.3.1 Transect Intercept Cover Sampling

Transect sampling – collecting data at defined points along a line – provides a relatively simple way to measure cover objectively. We present two different ways to measure shrub cover along a transect: point-intercept and line-intercept.

5.3.1.1 Point-Intercept Cover

The point-intercept method records shrub vegetation that intercepts or "hits" a pole perpendicular to a transect at specified sampling intervals (for example, hits along a 2 m pole placed at 1 m intervals) (Fig. 5.2). The optimal point-sampling interval and numbers of transects needed would depend upon vegetation density, and would be best determined by pilot testing. Hits at a given point can include only the uppermost layer or those in some specific layering scheme. To obtain the most flexible data for calculating cover layers or density measures, every vegetation hit can be recorded at each point.

Equipment

- 30 m tapes
- Pole (generally 2 m) marked in centimeter intervals
- Compass
- Chaining pins

Fig. 5.2 The point-intercept method tallies vegetation hits touching a pole perpendicular to a transect at predetermined intervals

Procedure

Using 30 m tapes, establish transects that radiate from each FIA subplot center. Offset shrub-measurement transects from deadwood transects, to minimize site disturbance and trampling that would affect other protocol measurements. Use pins to secure tapes at each end of the transect to keep spacing between sample points accurate. There is probably no need to slope-correct transects (because cover is not related to a strict map-area-based sampling frame); however, if slope correction is desired, then the total length of the slope-corrected transect would be divided by the number of sample points to obtain a new distance between points.

At each sample point along the transect, record either the highest intersection of vegetation hitting the pole or the heights of the highest intersections within height classes, if cover at different heights or layers of vegetation is desired. To obtain the most flexible data for calculating cover layers or density measures, one could also record multiple vegetation hits at a given point. Vegetation hits can be recorded by species or species group, depending on the desired detail.

The vertical pole should be perpendicular to the transect tape; a plumb-line might be useful to ensure field crew consistency. A vegetation hit needs to be carefully defined to account for irregular leaf and branching shapes and leaf movement on windy or rainy days.

Calculations

Percent shrub cover for a plot is calculated from point-intercept sampling by summing the number of sample points where vegetation intersects or hits the pole and dividing by the total number of points sampled. Cover can be calculated for each height layer if more than one layer is sampled.

$$\text{cover}_{jk} = \sum_{i=1}^{N} 100 \frac{P_{ijk}}{N} \tag{5.1}$$

where

cover_{jk} = percent shrub cover for height layer j of species or species group k

$$P_{ijk} \begin{cases} 1 \text{ if foliage intercepts sample pole at point i for height layer j of species k} \\ 0 \text{ otherwise} \end{cases}$$

N = total number of points sampled in layer j for all transects on plot

Note that one hit is generally used at each point (i) for any given cover or cover layer calculation. However, if more hits are recorded there is an opportunity to study the frequency of hits within height layers, which may be useful for constructing biomass equations from these data.

Biomass is calculated from equations using the calculated cover but these equations most likely will need to be developed. Measuring biomass for construction of these equations (as described later) will be somewhat complicated because two different sample frames will be needed since biomass cannot be estimated directly from transects. However, the point-intercept method has the flexibility of describing cover density in two dimensions (by using layers), which may prove quite useful in constructing generalized biomass equations for species groups.

Summary

This quantification of foliar frequency along transects provides an objective estimate of shrub cover at various layers for each species or species group. It is possible to use point-intercept sampling even in dense mixed-species communities where the foliage of many species overlaps. The point-intercept method is easy to implement without intensive training of field personnel. Its limitations for use in biomass estimation include: the possibility of bias in pole-intercept measurement for recording "hits" and the lack of appropriate species-level equations for converting cover to biomass. But the method offers much potential for constructing new biomass equations because of its flexibility to objectively measure shrub cover hits by species at multiple layer heights (which would provide additional variables related to density of shrub branching patterns, which in turn may help with biomass prediction).

5.3.1.2 Line-Intercept Cover

The line-intercept method records vegetation cover by measuring the length (horizontal distance) of shrub cover that intersects the transect (Fig. 5.3). Generally only the uppermost layer of cover is measured. The idea is to measure the extent of canopy that shades the ground when the sun is directly overhead.

Equipment

- 30 m tapes
- Pole (generally 2 m) marked in centimeter intervals
- Compass
- Chaining pins

Procedure

Using 30 m tapes, establish transects that radiate from each FIA subplot center. Offset shrub-measurement transects from deadwood transects, to minimize site disturbance and trampling that would affect other protocol measurements. Use pins to secure tapes at each end of the transect to keep spacing between sample points accurate. There is probably no need to slope-correct transects (because cover is not

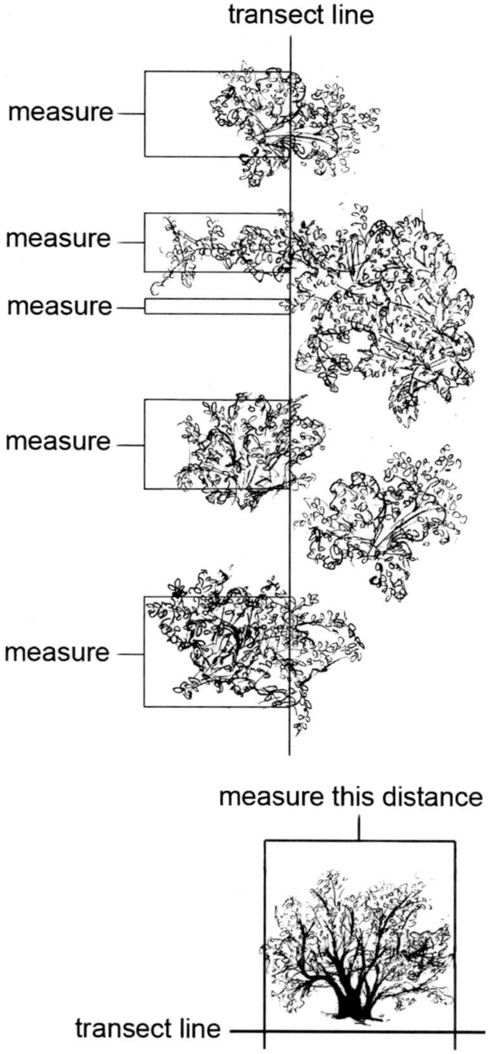

Fig. 5.3 Line-intercept methods measure the length of vegetation intersecting the transect on a horizontal plane

related to a strict map-area-based sampling frame); if slope correction is desired, then cover would be measured along the total length of the slope-corrected transect. However, calculations of slope-corrected transects would use horizontal transect length in the formula (which could result in more than 100% cover because a slope-corrected transect is longer than a horizontal transect).

Record vegetation that intersects the transect from an aerial vantage point (looking down from the top of the canopy). With the line-intercept method, generally only the uppermost cover layer is measured (because this usually is the widest part of the shrub), which can make species separation difficult when different species overlap at varying heights. Also, the actual measurement of vegetation length (horizontal distance) that intercepts the transect needs to be carefully defined to account for small open spaces, dead foliage, species overlap, and so forth. For example, a common definition to account for gaps is to count every space larger than 2 cm within the canopy as a gap.

Calculations

Percent shrub cover for a plot from line-intercept sampling is calculated by summing the total vegetation length intersected by all transects, then dividing by the total length of all transects sampled.

$$\text{cover}_k = \sum_{i=1}^{N} 100 \frac{d_{ik}}{L} \tag{5.2}$$

where
cover_k = percent shrub cover for species or species group k
d_{ik} = total distance or length of transect intercepted by species k (m)
N = total number of distance or intercept measurements for species k
L = total length (horizontal length, if slope corrected) of all transects on plot (m)

Biomass is calculated from equations using the calculated cover but these equations most likely will need to be developed. Measuring biomass for constructing these equations (as described later) will be somewhat complicated because two different area frames will be needed since biomass cannot be estimated directly from transects.

Summary

As with point-intercept, the line-intercept method also is objective and easy to teach to field personnel. The limitations of the line-intercept method are also similar to those of point-intercept: the lack of species-appropriate conversion equations for East Coast shrub species and the lack of a tested correlation between total estimated plot shrub cover and total plot biomass of shrub species. Furthermore, the line-intercept method may be too complicated for measuring individual species when many species overlap at different layers – particularly for eastern deciduous forests. On the other hand, it may work well for sparse shrub understory in dryland forests of the western US. Cover height can be added easily to obtain an additional variable for developing biomass equations from line-intercept cover data.

5.3.2 *Visual Cover Sampling*

Visual (or ocular) cover estimation within fixed-area plots is a relatively quick and popular way to estimate cover (Fig. 5.4). Cover can be visually estimated in units either to the nearest percent or within some predetermined cover classes (such as FIA's classes described earlier in this chapter). Although it is not possible to actually estimate cover to the nearest percent, recording the best estimate possible (to the nearest percent) is sometimes preferred over the use of cover classes when averaging many estimates. For one reason, large class boundaries may accentuate errors when dealing with border-line calls; furthermore, data collected in classes is problematic for averaging because the usual assumption of "class midpoint equal to class mean" can lead to bias when actual cover is not normally distributed within classes.

Two variations of this method are suggested that will fit within the FIA design: (a) cover estimates for microplots arranged within an FIA subplot or (b) cover for

Fig. 5.4 Visually estimating shrub cover is a simple and popular technique but requires careful attention for consistent application

small quadrats placed along transects within a subplot. By choosing either method, the FIA nested design is reduced to a single sampling frame, which is more consistent with our biomass objective.

5.3.2.1 Visual Cover for Microplots

The FIA design includes one 2.1 m radius microplot within each subplot, which is logical for sampling visual cover. However, the use of more microplots and different size radii might improve the estimate or sampling efficiency. We recommend that some pre-testing be conducted before deciding on the size and number of microplots needed.

Equipment

- 30 m tapes
- Compass
- Stakes or chaining pins
- Flagging
- Percentage-area guides for aiding visual estimation
- Pole (generally 2 m) marked in centimeter intervals

Procedure

Within each FIA subplot, locate and identify the microplot boundary with stakes or flagging. Estimate vegetation cover (of species or species group) from an aerial vantage point; if done by more than one person, average the estimates. For more repeatable and consistent estimation, it might be useful to divide the microplot into four quadrants; estimate the quadrants separately and then average the results. Estimation guides, such as pieces of cloth representing known quadrant areas, also are useful, at least for initial training. For example, each quadrant of a 2.1 m microplot would have square cloth guides with 59, 42, or 19 cm side dimensions to estimate 10%, 5%, or 1% of the quadrant area, respectively. Other estimation guides include pictures of microplots in shaded dot patterns representing percentages of cover for random and clumped arrangements for each quadrant. Cover height can also be measured with a pole, which provides an additional variable for cover-to-biomass modeling.

Calculations

No cover calculations (other than averages of all estimates within a plot) are needed because percent cover is estimated directly on the microplot or quadrant by species or species group. Two or more crew members also could make independent estimates of the same spots for additional averaging. Biomass is calculated from equations using the estimated cover. Some cover-based equations exist but more likely will need to be developed. Measuring biomass for constructing these equations (as described later) will be straightforward because shrubs on the microplots or quadrants simply can be cut and weighed.

Summary

Visual estimation is a popular, quick, and easy way to estimate cover but it requires careful training and quality control to maintain consistency. Forestry has a rich history of many such ocular measurements that can be quite effective under the right circumstances. The primary limitation is the subjectivity of visual estimates (there is often a credibility/objectivity worry about "visual estimates" depending on the conscientiousness of field crews), as well as a lack of accurate cover-to-biomass equations. Therefore, careful comparison testing against a more objective method is suggested before selecting visual cover estimation on microplots. Also, some type of estimation aid as described above is highly recommended.

5.3.2.2 Visual Cover Using Small Quadrats Spaced Along Transects

This variation of visual cover estimation uses small quadrats spaced systematically along transects. The method can fit well within the FIA design by using transects offset from the deadwood transects. The exact length and number of transects as well as the number and size of quadrats will depend on the density and species attributes of the vegetation being measured. Generally, this method is best for small shrubs measured in 1 m^2 (or smaller) quadrats. Size and numbers of quadrats should be determined in pre-testing. As an example, recommendations from a case study are presented later in this chapter.

Equipment

- 30 m tapes
- Compass
- Quadrat made of PVC pipe tubing
- Chaining pins
- Percentage-area guides for aiding visual estimation
- Pole (generally 2 m) marked in centimeter intervals

Procedure

Using 30 m tapes, establish several transects that radiate from each FIA subplot center, offset from deadwood transects (to avoid over trampling). Place one edge of the quadrat over the tape at fixed intervals to visually estimate cover. There is probably no need to slope-correct transects (because cover is not related to a strict map-area-based sampling frame); however, if slope correction is desired, then the total length of the slope-corrected transect would be divided by the number of sample quadrats to obtain a new distance between quadrats.

Estimate vegetation cover to the nearest 1% (for each species or species group) from an aerial vantage point; if done by more than one person, average the results. For more repeatable and consistent visual estimation, it might be useful to divide the quadrat into four sections; estimate the sections separately and then average the results. Cloth estimation guides representing known quadrat areas, or a mylar sheet on which quadrats with known percentages are drawn, also are useful, at least for initial training. For example, square cloth guides or darkened quadrats on a mylar sheet for a 1 m^2 quadrat would have 32, 22, or 10 cm side dimensions to estimate 10%, 5%, or 1% quadrat area, respectively. Even simpler, a person's fist is about 10 cm^2, or 1% of a 1 m^2 quadrat (Fig. 5.5). Cover height can also be estimated for an additional variable that might be useful for cover-to-biomass modeling.

Calculations

No formal cover calculations are needed because percent cover is estimated directly on quadrats by species or species group. However, two or more crew members could make independent estimates that are averaged. Like the microplot variant,

Fig. 5.5 An average person's fist is about 10 cm^2, or 1% of a 1 m^2 quadrat

biomass is calculated from equations using the estimated cover. Some cover-based equations exist but more likely will need to be developed. It will be a little more complicated to measure biomass for many small quadrats than for fewer and larger microplots because edge effects will be more pronounced for quadrat estimates. Since the quadrat cover-to-biomass relationship is defined by only that cover within a quadrat, careful discarding of all branches outside the quadrat will need to be done before quadrat cover is cut for weighing.

Summary

This quadrat method offers more opportunity to spread out small samples for visual cover than does the microplot variant. Using a smaller sampling area allows for greater accuracy and consistency, especially with training and estimation aids. However, it will take some preliminary work to establish appropriate quadrat size, which should probably be larger than crown spread of largest shrub size. Also, choice of appropriate quadrat size can be further complicated when large and small shrubs are present on the same plot. Like the microplot variant, the primary limitation of the quadrat approach is the subjectivity of the cover estimation; however, sampling and averaging many small quadrats provides greater opportunity to minimize potential bias, if the results both overestimate and underestimate true cover. Also, like the other cover estimation methods, there is a lack of accurate cover-to-biomass equations; these should be developed before using any method of cover estimation to represent forest shrub biomass. Nevertheless, the use of small quadrats is among the easiest ways to estimate percent cover.

5.3.3 Diameter Measurement (BA)

The final method also uses a fixed-area microplot but, in this case, the basal diameter of each shrub stem is measured. This is the most objective method, offering much flexibility for long-term monitoring of shrub growth and mortality. Because shrub growth is part of net primary production for understory – which can be comparable to that of trees (Nilsson and Wardle 2005) – it might be worth the extra effort to measure growth of shrub stems. However, dense mats of small-stemmed shrubs such as *Vaccinium* species can be particularly time-consuming to measure. Nevertheless, the shrub species definition can be modified to add a minimum diameter (5 or 10 mm, for example) below which the species would be treated as an herb and measured with a more appropriate cover-based procedure. Although there is some disagreement in the literature on where to measure shrub diameter (Ohmann et al. 1976), we advocate measuring each stem near the root collar (drc) but just above any abnormal swell (Fig. 5.6).

If a total shrub plant diameter is desired, the equivalent diameter (edrc) is recommended (Chojnacky 1994):

$$\text{edrc} = \sqrt{\sum_{i=1}^{n} \text{drc}_i^2} \tag{5.3}$$

where
n = number of stems measured for drc in individual shrub plant

Since FIA protocol includes one 2.1 m radius microplot within each subplot, shrubs could be measured here along with tree saplings. However, adding more microplots per subplot may increase the accuracy of the estimation. The size and number of the microplots could vary depending on the species mixture in different parts of the US. In our experience in the eastern US, the FIA microplot seems a little larger than necessary, for example. Several smaller quadrats within subplots might also be tested to better capture the spatial variation of shrub communities.

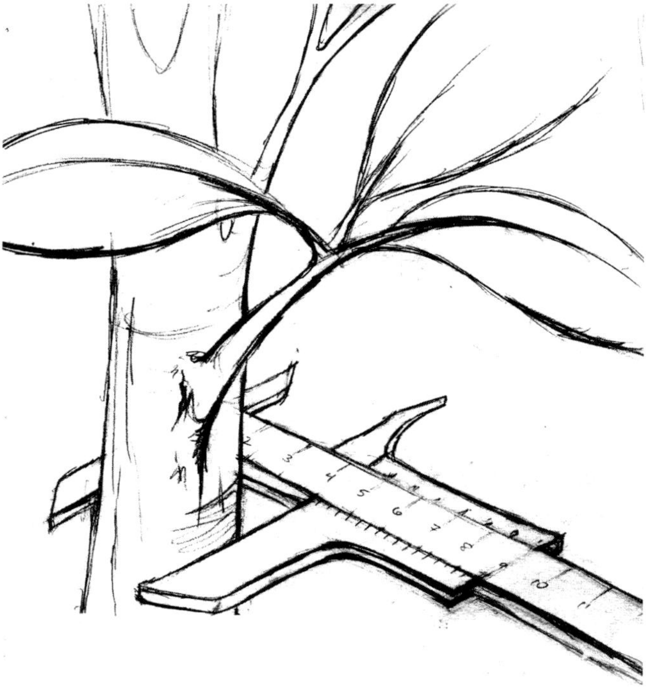

Fig. 5.6 Diameter measurement is the most objective method but can require considerable effort

Equipment

- 30 m tapes
- Caliper
- Diameter tape

- Pole (generally 2 m) marked in centimeter intervals
- Compass
- Stakes or chaining pins
- Flagging

Procedure

On microplots, measure the drc of each shrub stem larger than some minimum diameter (probably 5 or 10 mm). For permanent plot installations, the center point of the microplot can be marked and the distance and azimuth to some or all shrub stems can be recorded for future relocation and remeasurement. In addition to species, stem status (live or dead) can be recorded to assess mortality and dead material. Height of some or all stems can also be recorded for potential regression variables for biomass estimation.

Minimums matter
The reason for choosing a minimum diameter is that some shrubs—such as *Vaccinium*—can grow in low dense clumps with many stems smaller than 10 mm in diameter. It is probably more reasonable to include shrub stems less than 5, 8, or 10 mm in diameter with herbaceous cover measurements rather than with shrubs. Furthermore, the comparatively little carbon accounted for by small shrubs does not warrant the exponential increase in field work required to measure all small shrubs.

Calculations

For these data, shrub frequency, basal area, and biomass per hectare can be calculated for live and dead material in size classes. Biomass, of course, requires an appropriate drc-based equation, or one needs to constructed as described below. For illustration, we show calculation of total basal area:

$$Y_k = \sum_{i=1}^{S_k} \frac{\pi \, drc_{ik}^2}{40{,}000 \, A_{mp}} \tag{5.4}$$

where
Y_k = total basal area for species k (m^2/ha)
drc_{ik} = diameter near root collar for stem i of species k (cm)
A_{mp} = area of microplot (ha)
S_k = total number of stems for species k measured on microplot

Similarly, biomass could be computed with an appropriate diameter-based regression equation. Ideally, regression equations are desired for each species or for a group of similar species for different sections or areas of the U.S. However, recent theoretical work on allometric scaling (as will be discussed later) may simplify numbers of equations needed.

Summary

Diameter measurements are the most time-intensive methods presented, yet they are also the most objective, accurate, and flexible measurements that can be made for calculating shrub biomass. Such measurements can easily be combined with the FIA measurement of seedlings and saplings. The limitation of the diameter measurement method is the cost of these time-consuming measurements, particularly for dense shrub cover. However, this drawback might be balanced with carefully crafted plot sizes to minimize unnecessary measurement effort.

5.4 Summary of Recommended Methods

Table 5.1 provides a summary of the methods presented in this chapter, including strengths and limitations. For details on recommended sample sizes and dimensions for each method, consult the appropriate references.

5.5 Regression Subsampling for Carbon Estimation

The three methods described above will not estimate carbon directly without an appropriate regression equation. The best way to develop a suitable regression equation is to subsample some of the plots (or nearby plots if they are permanent) to measure actual shrub weights and then use the subsamples to develop regression equations from actual cover or diameter measurements.

Developing appropriate biomass equations is particularly important for the cover measurements (both transect and visual cover methods), because cover-to-biomass relationships can vary widely depending on the height and branching pattern of the shrub species. Therefore, considerable original research likely will be needed to develop widely applicable equations. For example, the point-intercept method might be scaled to a broad range of species by using cover (or point-intercept) measurements at various heights as variables in the equations. Likewise, visual cover estimation methods might benefit from collecting cover height variables to develop more general multiple species equations. Because only a few cover-to-weight equations exist (Olson and Martin 1981, Humphrey 1985,

Table 5.1 Comparison of shrub biomass measurement and estimation methods

Method	**1-Transect intercept cover sampling**		**2-Visual cover sampling**		**3- Diameter measurement sampling**
Sample frame	Transect	Transect	Fixed-area microplot	Fixed-area quadrat	Fixed-area microplot
Variable measured	Cover	Cover	Cover	Cover	Diameter
Method variant	1a-Point-intercept	1b-Line-intercept	2a-Microplot	2b-Quadrat	None
Description	Vertical vegetation canopy points (hits) intercepting perpendicular pole are tallied along transect	Horizontal vegetation canopy length intercepting transect is measured	Overhead vegetation cover is visually estimated on microplot	Overhead vegetation cover is visually estimated on small quadrats systematically spaced along one or more transects	Shrub stem diameters larger than some threshold are measured at drc on microplot
Sample size variables	Transect length and numbers; sample points per transect	Transect length and numbers	Microplot area and number per subplot	Quadrat area and number per transect; transects per subplot	Microplot area and number per subplot
Biomass calculations	First, cover calculated from no. of sample points/total points; then biomass-to-cover equation needed	First, cover calculated from total length of vegetation intersected/total length of all transects; then biomass-to-cover equation needed	Cover estimated directly but biomass-to-cover equation needed.	Cover estimated directly but biomass-to-cover equation needed	Biomass-to-diameter equation needed
Strengths	Objective; rapid; easy to separate species and cover layer; flexibility from height layer variables for constructing new biomass equations	Objective; rapid; easy to teach; can easily get added height variable for constructing new biomass equations	Most rapid; easy; widely used	Rapid; opportunity to statistically assess many small samples	Objective; most accurate and suitable for use in biomass equations; flexible for other long-term monitoring needs

(continued)

Table 5.1 (continued)

Method	**1-Transect intercept cover sampling**		**2-Visual cover sampling**		**3- Diameter measurement sampling**
Sample frame	Transect	Transect	Fixed-area microplot	Fixed-area quadrat	Fixed-area microplot
Variable measured	Cover	Cover	Cover	Cover	Diameter
Method variant	1a-Point-intercept	1b-Line-intercept	2a-Microplot	2b-Quadrat	None
Limitations	Cumbersome for dense mixed species communities; precaution needed to avoid pole-intercept measurement bias; likely need new biomass equations, and development not straight–forward	Difficult to separate species in dense mixed communities; likely need new biomass equations, and development not straight–forward	Subjective; precaution needed to minimize personal estimation bias—particularly for large microplots; lack of cover-to-biomass equations	Subjective; precaution needed to minimize personal estimation bias; lack of cover-to-biomass equations	Most time-consuming and costly method; biomass equations lacking for many species
References*	Diersing and others 1992; Elzinga and others 1998; Heady and others 1959; Korb and others 2003	Canfield 1941; Elzinga and others 1998; Hanley 1978; Sweeney and Cook 2001	FIA 2002; Bonham 1989, p.101; Daubenmire 1959; Humphrey 1985; Small and McCarthy 2002	Elzinga and others 1998; FIA 2002	Brown 1976; Chojnacky 1984; Ohmann and others 1976; Tefler 1969

*The major reference for each method is listed first.

Mitchell et al. 1987, Gilliam and Turrill 1993, Means et al. 1994), considerable effort on subsampling should be planned until a pool of equations becomes available.

Developing these relationships for the transect methods will take extra thought, because transect point- and line-intercept methods cannot estimate biomass directly. Instead, two sampling frames likely are needed where biomass is estimated from fixed-areas, which will need to be related to cover measured on transects for the same areas. For example, cover that is measured on parallel transects or other patterns of transects can be matched to biomass that is estimated for the same area. Measurement of shrub biomass for large areas will likely be too costly; hence, microplots or quadrats would be needed to subsample biomass for the entire area sampled by transects. Rectangular belts around each transect could also be considered for estimating biomass. For certain conditions, there is a transect method pioneered by Meeuwig and Budy (1981) for estimating biomass of shrubby pinyon-juniper trees; this method might be useful for developing shrub biomass equations. We did not present Meeuwig and Budy's method here for shrub carbon estimation because their method requires simple plant communities where individual crown diameters can be easily identified; however, it might be useful for verifying cover-to-biomass regressions for transect methods.

Basal (diameter) area (BA) measurement is probably the most robust method for grouping many species into a few biomass equations, because diameter-to-weight is a strong empirical relationship (Yandle and Wiant 1981). Also, diameter-to-weight relationships have a theoretical basis in allometric scaling theory (Enquist and Niklas 2001, 2002) (see sidebar). However, shrub biomass equations do not exist for all species; some are scattered throughout the forestry, range, and ecology literature (Tefler 1969, Brown 1976, Brown and Marsden 1976, Ohmann et al. 1976, Alaback 1986, Elliott and Clinton 1993). A comprehensive summary is included in BIOPAK software (Means et al. 1994) for Pacific Northwest species. As recommended for cover measurement, considerable effort on subsampling should be planned until more diameter-to-biomass equations become available.

Allometric theory supports basal measurement method

The scientists who developed scaling theory took clues from naturally occurring networks. For example, the lung circulatory system and tree or shrub branching follow similar patterns. The theory shows promise for tree biomass estimation even for multiple-stemmed trees (Chojnacky 2002), which implies it may also be worthwhile for shrubs (Adapted from *New York Times* 01/12/1999).

(continued)

Allometric theory supports basal measurement method (continued)

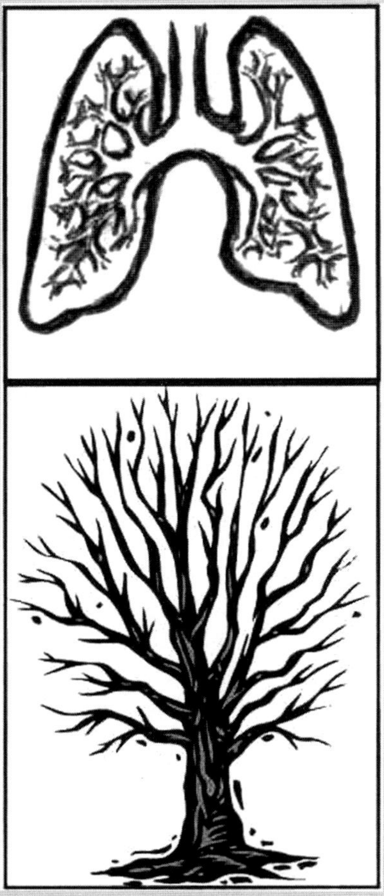

5.6 Biomass Measurements

Biomass measurement for constructing equations can be done by destructively cutting and weighing entire shrub plants or by using various subsampling techniques to cut and weigh plant portions (Bonham 1989, p. 228; Gregoire et al. 1995). If a subsampling technique is used, we recommend that the subsampling scheme be tested against entire plant harvesting to document the method's effectiveness. If a subsample of plots is routinely harvested as part of the sample design (by using classical regression sampling), we still recommend that auxiliary regression equations for the weight data be published. This will speed the daunting task of developing good weight relationships for all shrub species. As more shared data become available, perhaps allometric scaling or other theories can be applied to speed the process of studying one species at time at a few specific sites. Also, if biomass data are collected, it would be useful to check the commonly held assumption that carbon is indeed 50% of dry biomass for all species.

5.7 Applying the Equations: An Example

An exhaustive meta-analysis of shrub biomass equations is beyond the scope of this chapter but examples are given to illustrate how our methods might be applied to estimate carbon in shrubs from cover or diameter measurements. Each particular application will require careful consideration of equations selected; however, examples from two existing studies are provided as a conceptual stepping stone for developing specific methodology that is appropriate to varying situations in different places.

Ideally, both the equations and the data developed would come from the same section or region of the US, unlike our situation in which the equations were developed for West Coast species and the data were collected on East Coast shrub species. Therefore, our specific results are less the focus for this chapter than the demonstration of the methods.

For our example, the equations in BIOPAK (Means et al. 1994) were summarized in a meta-analysis (as done for trees in Jenkins et al. 2003) into two equations for either cover or basal diameter measurement:

$$\begin{aligned} \text{biomass}_T &= Exp\,[-3.96457 + 1.08631 \ln(\text{cover})] \\ \text{biomass}_P &= Exp\,[-3.42620 + 2.5031 \ln(\text{drc})] \end{aligned} \tag{5.5}$$

where
biomass_T = total shrub dry weight (Mg/ha)
biomass_P = individual shrub stem or plant dry weight (kg/plant)
cover = percent shrub cover
ln = natural logarithm
drc = basal diameter of each shrub stem near root collar (cm)

These equations were then applied to nine vegetation plots measured for both cover and drc in eastern hardwood forests in national parks near Washington, DC. Primary species included *Kalmia latifolia, Lindera benzoin, Vaccinium* species, and *Viburnum* species. The point-intercept method was used to measure cover up to 2 m in height on three 18 m transects sampled at 1 m intervals. Basal stem diameters (drc) were measured on three 4 m radius microplots within the same area for all shrub stems 5 mm and larger. The sample area for both methods was an 18 m circular plot where the transects radiated out from the center at 120° angles. Individual biomass measurements were summed and appropriately expressed (using sample weights) to arrive at biomass per hectare.

Results of the test are informative and illustrate some of what to bear in mind when estimating shrub carbon (Table 5.2). First, the large differences in biomass between the cover/transect and diameter/plot sampling methods are likely due to the inaccuracy of West Coast auxiliary equations for East Coast shrub species. The cover

Table 5.2 Comparison of two methods for estimating shrub biomass from auxiliary equations applied to diameter and cover measurements made on plots in national parks near Washington, DC

Plot information			Drc plot measurement-summary		Point-intercept measurement summary		Converted biomass*	
			Stems	Qmd	Cover	Height	Point-intercept-cover	Diameter measurement (drc)
No.	Shrub Type	Park	*No./ha*	*cm*	*%*	*m*	- - - -*Mg/ha*- - -	
P01	Kallat	Prince William, VA	1,658	3.9	9	0.9	0.2	15.8
S01	Kallat	Rock Creek, DC	6,565	2.2	26	1.1	0.7	15.3
N04	Linben/ Vac/Vib	Rock Creek, DC	20,961	1.1	74	1.3	2.0	10.7
T02	Linben/ Vac/Vib	Catoctin, MD	39,689	0.7	59	0.2	1.6	10.3
T01	Kallat	Catoctin, MD	531	2.8	0	0.0	0.0	2.1
C01	Linben/Vac/ Vib	Rock Creek, DC	3,714	0.8	20	0.6	0.5	0.6
C06	Linben/Vac/ Vib	Rock Creek, DC	1,857	0.8	4	0.2	0.1	0.5
T03	Linben/Vac/ Vib	Catoctin, MD	3,714	0.5	2	0.3	0.0	0.3
C03	Linben/ Vac/Vib	Rock Creek, DC	66	0.8	0	0.0	0.0	0.0

Species codes: Kallat–(*Kalmia latifolia*), Linben–(*Lindera benzoin*), Vac–(*Vaccinium* species), Vib–(*Viburnum* species); Abbreviations: qmd=quadratic mean stem diameter, drc=diameter at root collar, cover=upper layer cover, height=height of highest layer; * Individual measurements were summed and converted to total forest dry weight using equation 5.5 and appropriate area expansion.

equation is from a conservative meta-analysis that combines low ground shrubs with some taller shrubs; it limits predictions to less than 3 Mg/ha no matter how much cover is actually present. Had height of cover layer or additional cover layers been included, the equation would have been more flexible. On the other hand, the diameter-based (drc) equation is probably more robust (and realistic), producing a larger and wider range of results. Therefore, these results illustrate that appropriate and realistic biomass equations are key to good carbon estimates, regardless of which method is used. Although developing good cover-to-biomass regressions will be quite difficult, costly, and time-consuming, the positive trade-off is that good equations will enable the use of rapid cover measurements from that point on.

Although not discernible in Table 5.2, the test also revealed that the largest shrub, mountain laurel (Kallat), accounted for considerable difference between the two methods. For the sake of illustration, if mountain laurel is left out of the

Table 5.3 Comparison of two methods for estimating shrub biomass from auxiliary equations, with mountain laurel (*Kalmia latifolia*) dropped from the analysis

Plot information		Converted biomass*	
		Point-intercept cover	Diameter measurement (drc)
Number	Park	------- *Mg/ha* ------	
P01	Prince William, VA	0.1	0.5
S01	Rock Creek, DC	0.5	0.6
N04	Rock Creek, DC	2.0	10.7
T02	Catoctin, MD	1.6	0.9
T01	Catoctin, MD	0.0	0.0
C01	Rock Creek, DC	0.5	0.6
C06	Rock Creek, DC	0.1	0.5
T03	Catoctin, MD	0.0	0.3
C03	Rock Creek, DC	0.0	0.0

* Individual measurements were summed and converted to total forest dry weight using equation 5.5 and appropriate area expansion.

previous analysis, the diameter measurement and cover methods are more comparable (Table 5.3) because there is less mixing of high and low shrubs. However, even here the largest difference in estimated biomass (plot N04) can again be attributed to many small stems of spice bush (*Lindera*) that add up to more biomass than would be obtained from transect cover conversion.

While there is no way of stating here for certain which method will be best for a particular project, the diameter measurement method generally will be the most sensitive because actual diameters tend to be directly proportional to biomass; thus our preference for diameter measurement (drc) as the most robust method. However, the cover methods also can be used to provide reasonable estimates if some work is done to develop separate equations for individual species groups; inclusion of height and/or more cover layers also might make the equations more robust.

5.8 Summary

A reasonable estimate of carbon in shrubs can be achieved by combining general cover or diameter measurements (using transects or microplots arranged within FIA subplots) with regression subsampling. Stem diameter measurements using microplots seem to provide the most robust data for estimating biomass (and by extension, carbon). However, these measurements require the most field time; diameter-to-biomass equations are lacking for many species, which will require considerable research work for wide application of this method; and more pilot study should be done to determine optimum size and number of microplots.

More rapid measurements of cover also can be used when appropriate, realistic cover-to-biomass equations are developed. The point-intercept method – which inherently gives cover layer and height data – might be most advantageous for developing robust cover-to-biomass relationships. This method offers a rapid but objective procedure for estimating shrub cover for individual species by layers. Although corresponding biomass equations must be developed, there seems to be considerable opportunity to develop such equations from a range of variables including cover by species at a variety of layer heights.

The need to accurately estimate carbon in shrubs will become even more vital in the future as land use and global climate changes increasingly alter both forested and non-forested ecosystems. Such changes will have implications for wildlife, biodiversity, nutrient cycling, fire, and other management issues where carbon sequestration, especially in soils, may become key objectives. We have summarized three methods that might be used to estimate carbon in shrubs. The job now is to develop appropriate equations and workable field measurement techniques that together will enable researchers and managers to measure carbon in shrubs quickly, consistently, and rigorously in the future.

Acknowledgements We gratefully acknowledge the assistance of the National Capital Region of the U. S. Department of the Interior, National Park Service, for sharing data used in example calculations; the Northern Global Change Program of the U.S. Department of Agriculture, Forest Service, for financial support; Mary Carr of the Forest Service, CAT Publishing Arts, for editing the manuscript; Michal J. Chojnacky of George Washington University for the illustrations; and Kristen Thrall of the Forest Service, Recreation Solutions, for graphics assistance.

Literature Cited

Alaback PB (1986) Biomass regression equations for understory plants in coastal Alaska: effects of species and sampling design on estimates. Northwest Science 60(2): 90–102

Bonham CD (1989) Measurements for terrestrial vegetation. New York: Wiley, 354 p

Bechtold WA, Patterson PA (Eds.) (2005) The enhanced Forest Inventory and Analysis program—national sampling design and estimation procedures. Gen. Tech. Rep. SRS-80. Asheville, NC: U.S. Department of Agriculture, Forest Service, Southern Research Station, 85 p

Brown JK (1976) Estimating shrub biomass from basal stem diameters. Canadian Journal of Forest Research 6: 153–158

Brown JK, Marsden MA (1976) Estimating fuel weights of grasses, forbs, and small woody plants. Res. Note. INT-210. Ogden, UT: U.S. Department of Agriculture, Forest Service, Intermountain Forest and Range Experiment Station, 11 p. www.fs.fed.us/rm/main/pubs/inventories/int/int_rn/pdf

Canfield R (1941) Application of the line-interception method in sampling range vegetation. Journal of Forestry 39: 388–394

Chojnacky DC (1994) Volume equations for New Mexico's pinyon-juniper dryland forests. Research Paper INT-471. Ogden, UT: U.S. Department of Agriculture, Forest Service, Intermountain Research Experiment Station, 10 p

Chojnacky DC (1998) Double sampling for stratification: a forest inventory application in the Interior West. Res. Pap. RMRS-RP-7. Ogden, UT: U.S. Department of Agriculture, Forest Service, Rocky Mountain Research Station, 15 p

Chojnacky DC (2002) Allometric scaling theory applied to FIA biomass estimation. In: McRoberts RE, Reams GA, Van Deusen PC, Moser JW (Eds.) Proceedings of third annual forest inventory and analysis symposium, 17–19 October 2001, Traverse City, MI. Gen. Tech. Rep. NC-230. St. Paul, MN: U.S. Department of Agriculture, Forest Service, North Central Research Station, pp. 96–102

Cochran WG (1977) Sampling Techniques. Third edn. New York: Wiley, 428 p
Daubenmire RF (1959) Canopy coverage method of vegetation analysis. Northwest Science 33: 43–64
Diersing VE, Shaw RB, Tazik DJ (1992) U.S. Army land condition-trend analysis (LCTA) program. Environmental Management 16: 405–414
Elliott KJ, Clinton BD (1993) Equations for estimating biomass of herbaceous and woody vegetation in early-successional southern Appalachian pine-hardwood forests. Res. Note SE-365. Asheville, NC: U.S. Department of Agriculture, Forest Service, Southeastern Forest Experiment Station, 7 p
Elzinga, CL, Salzer DW, Willoughby JW (1998) Measuring and monitoring plant populations. BLM Technical Reference 1730-1. Denver, CO: Bureau of Land Management, 477 p
Enquist BJ, Niklas KJ (2001) Invariant scaling relations across tree-dominated communities. Nature 410: 655–660
Enquist BJ, Niklas KJ (2002) Global allocation rules for biomass partitioning in seed plants. Science 295: 1517–1520
Forest Inventory and Analysis [FIA] (2002) Phase 3 field manual—vegetation diversity and structure, March 2002. U.S. Department of Agriculture, Forest Service, Forest Inventory and Analysis program, 20 p. http://fia.fs.fed.us/library/field-guides-methods-proc/docs/p3sec13_02-03_pub.pdf
Forest Inventory and Analysis (FIA) (2004a) FIA tree species codes. In: National core field guide, volume 1: field data collection procedures for phase 2 plots, appendix 3, January 2004. U.S. Department of Agriculture, Forest Service, Forest Inventory and Analysis Program, pp. 151–163 http://fia.fs.fed.us/library/field-guides-methods-proc/docs/core_ver_2-0_01_04.pdf [accessed June 21, 2005]
Forest Inventory and Analysis (FIA) (2004b) Phase 3 field manual—down woody materials, March 2004. U.S. Department of Agriculture, Forest Service, Forest Inventory and Analysis program, 38 p. http://www.fia.fs.fed.us/library/field-guides-methods-proc/docs/p3_2-0_sec14_3_04.pdf [accessed December 1, 2005]
Forest Inventory and Analysis (FIA) (2005) Homepage of U.S. Department of Agriculture, Forest Service, Forest Inventory and Analysis Program. [accessed June 14, 2005]
Gilliam FS, Turrill NL (1993) Herbaceous layer cover and biomass in a young versus a mature stand of a central Appalachian hardwood forest. Bulletin Torrey Botanical Club 120: 445–450
Gregoire TG, Valentine HT, Furnival GM (1995) Sampling methods to estimate foliage and other characteristics of individual trees. Ecology 76(4): 1181–1194
Hanley TA (1978) A comparison of the line-interception and quadrat estimation methods for determining shrub canopy coverage. Journal of Range Management 31(1): 60–62
Heady HF, Gibbens RP, Powell RW (1959) A comparison of the charting, line intercept, and line point methods of sampling shrub types of vegetation. Journal of Range Management 12(4): 180–188
Humphrey DL (1985) Use of biomass predicted by regression from cover estimates to compare vegetational similarity of sagebrush-grass sites. Great Basin Naturalist 45(1): 94–98
Jenkins JC, Chojnacky DC, Heath LS, Birdsey RA (2003) National-scale biomass estimation for United States tree species. Forest Science 49(1): 12–35
Kimble JM, Heath LS, Birdsey RA (2002) The potential of U.S. forest soils to sequester carbon and mitigate the greenhouse effect. Washington, DC: CRC Press, 429 p
Korb, JE, Covington WW, Fule PZ (2003) Sampling techniques influence understory plant trajectories after restoration: an example from ponderosa pine restoration. Restoration Ecology 11(4): 504–515
Kort J, Turnock B (2003) Biomass production and carbon fixation by prairie shelterbelts: a green plan project. Agriculture and Agri-Food Canada, Supplementary Report 96-5 http://www.agr.gc.ca/pfra/shelterbelt/shbpub63.htm [accessed June 21, 2005]
Lohr SL (1999) Sampling: design and analysis. Pacific Grove, CA: Duxbury Press, 494 p
McMillin JD, Allen KK (2003) Effects of Douglas-fir beetle (Coleoptera: Scolytidae) infestations on forest overstory and understory conditions in western Wyoming. Western North American Naturalist 63(4): 498–506

Means J, Hansen H, Koerper G, Alaback P, Klopsch M (1994) Software for computing plant biomass- BIOPAK users guide. Gen. Tech. Rep. PNW-GTR-340. Portland, OR: U.S. Department of Agriculture, Forest Service, Pacific Northwest Research Station

Meeuwig RO, Budy JD (1981) Point- and line-intersect sampling in pinyon and Utah juniper. Paper INT-104. Ogden, UT: U.S. Department of Agriculture, Forest Service, Intermountain Forest and Range Experiment Station, 38 p

Mitchell, JE, Bartling PNS, O'Brien RO (1987) Understory cover-biomass relationships in the Front Range ponderosa pine zone. Research Note RM-471. Fort Collins, CO: U.S. Department of Agriculture, Forest Service, Rocky Mountain Forest and Range Experiment Station. 5 p www.fs.fed.us/rm/main/pubs/inventories/rm_rn.pdf.

Nilsson MC, Wardle DA (2005) Understory vegetation as a forest ecosystem driver: evidence from the northern Swedish boreal forest. Frontiers in Ecology and the Environment. 8(30): 421–428

Ohmann LF, Grigal DF, Brander RB (1976) Biomass estimation for five shrubs from northeastern Minnesota. Journal Series 9468. St. Paul, MN: University of Minnesota. Minnesota Agriculture Experimentation Station, 11 p

Olson CM, Martin RE (1981) Estimating biomass of shrubs and forbs in central Washington Douglas-fir stands. Res. Note PNW-380. Portland, OR: U.S. Department of Agriculture, Forest Service, Pacific Northwest Research Station, 6 p

Peet RK, Wentworth TR, White PS (1998) A flexible multipurpose method for recording vegetation composition and structure. Castanea 63: 262–274

Sweeney BA, Cook JE (2001) A landscape-level assessment of understory diversity in upland forest of the North-Central Wisconsin, USA. Landscape Ecology 16: 55–69

Small CJ, McCarthy BC (2002) Spatial and temporal variability of herbaceous vegetation in an eastern deciduous forest. Plant Ecology 164: 37–48

Tefler ES (1969) Weight-diameter relationships for 22 woody plant species. Canadian Journal of Botany 47: 1851–1855

Yandle DO, Wiant HV (1981) Estimation of plant biomass based on the allometric equation. Canadian Journal Forestry Research 11: 833–834

Chapter 6
Estimating the Carbon in Coarse Woody Debris with Perpendicular Distance Sampling

Harry T. Valentine, Jeffrey H. Gove, Mark J. Ducey, Timothy G. Gregoire, and Michael S. Williams

Abstract Perpendicular distance sampling (PDS) is a design for sampling the population of pieces of coarse woody debris (logs) in a forested tract. In application, logs are selected at sample points with probability proportional to volume. Consequently, aggregate log volume per unit land area can be estimated from tallies of logs at sample points. In this chapter we provide protocols and formulae for estimating the carbon in coarse woody debris with PDS. We also provide formulae for estimating components of change in the log population between two points in time.

Keywords Components of change, limiting distance, logs, two-phase sampling, volume factor

---

H.T. Valentine
U.S. Forest Service, Northern Research Station, 271 Mast Rd, Durham, NH 03824
E-mail: hvalentine@fs.fed.us

J.H. Gove
U.S. Forest Service, 271 Mast Rd, Durham, NH 03824
E-mail: jgove@fs.fed.us

M.J. Ducey
Department of Natural Resources, Univ of New Hampshire, 215 James Hall, Durham, NH 03824
E-mail: mjducey@unh.edu

T.G. Gregoire
School of Forestry and Environmental Studies, Yale Univ, 360 Prospect St, New Haven, CT 06511
E-mail: timothy.gregoire@yale.edu

M.S. Williams
Food Safety Inspection Service, 2150 Centre Ave, Building D, Fort Collins, CO 80526
E-mail: mike.williams@fsis.usda.gov

C.M. Hoover (ed.) *Field Measurements for Forest Carbon Monitoring*,

6.1 Introduction

Strategies for the estimation of attributes of fallen coarse woody debris on forested tracts have employed a variety of sampling designs, The most popular designs are line intersect sampling (e.g., de Vries 1986, Affleck et al. 2005) and plot sampling (e.g., Gregoire and Valentine 2008, Chap. 7), though several recently conceived designs – transect relascope sampling (Ståhl 1998), point relascope sampling (Gove et al. 1999), a prism sweep method (Bebber and Thomas 2003), and perpendicular distance sampling (Williams and Gove 2003, Williams et al. 2005) – also are applicable.

Coarse woody debris comprises the woody remains of fallen trees and large branches. Customarily, individual pieces of coarse woody debris are called 'logs.' In line intersect sampling, the probability that a particular log is included in the sample is proportional to log length. In biological organisms, plant or animal, length tends to scale with the 1/4 power of volume (e.g., West et al. 1999). Volume, in turn, tends to scale isometrically with carbon stock, our parameter of interest. In fixed-radius plot sampling or quadrat sampling, all logs in a sample are included with equal probability regardless of volume, so plot sampling is, in effect, sampling with probability proportional to the zeroth power of volume, which most likely is less efficient than line intersect sampling for estimating carbon stock. By contrast, in point relascope sampling, a log is included in a sample with probability proportional to the log's length-square, which tends to scale with the 1/2 power (or square root) of volume, so on theoretical grounds we should expect point relascope sampling to be more efficient than line intersect sampling, and this expectation is supported by results from field tests (Brissette et al. 2003). Ideally, the inclusion probability of a log should scale directly with its volume, and this is achieved with perpendicular distance sampling (PDS). Thus, in this chapter, we describe how the carbon in coarse woody debris can be estimated with PDS.

6.1.1 Coarse Woody Debris

Our objective is the estimation of areal carbon density for the population of pieces of coarse woody debris on the floor of a forested tract. By areal carbon density, we mean the carbon stock per unit land area (kg C ha^{-1}). The question at the outset is: what constitutes a piece of coarse woody debris?

We define any fallen stem and all its connected branches and sub-branches to be a single piece of woody debris. The debris is 'coarse' if its diameter meets or exceeds the minimal diameter (0.1 m = 10 cm) at some point on the main stem. Thus, under this protocol, every piece of wood on the ground, whether branched or not, is either coarse woody debris or fine woody debris. This definition is applicable to all sampling designs. However, since fallen wood gradually decays, we also need a way of distinguishing decomposed coarse woody debris from organic soil. We shall adopt a

Table 6.1 Decay classes of logs

Class	Integrity	Texture
1	Sound, freshly fallen	Intact, no rot
2	Sound	Intact, sapwood partly soft
3	Heartwood sound, log supports its own weight	Sapwood can be pulled apart by hand or sapwood is absent
4	Heartwood rotten, log does not support its own weight, but maintains its shape	Soft, small blocky pieces; a metal pin can be pushed into heartwood
5	None; log spreads on the ground	Soft, powdery when dry

decay classification used by the U.S. Forest Service (Table 6.1). Decay classes 1–4 connote wood; decay class 5 is organic soil.

For ease of presentation, we shall henceforth follow custom and call all pieces of coarse woody debris 'logs.' Moreover, unless otherwise specified, all log lengths and diameters are measured in meters; cross-sectional areas of logs are measured in square meters; volume is measured in cubic meters, and carbon mass is measured in kilograms. With one exception, land area is measured in ha. A symbol table is provided in the appendix (Table 6.4).

6.2 Perpendicular Distance Sampling

We presume that an investigator selects the locations of m sample points, P_s, $s = 1, 2, ..., m$ within the tract of interest. These points may be selected independently at random, but ordinarily the sample points are arranged either singly or in clusters on a systematic grid, where the anchor point of the grid is selected independently at random. The acceptance-rejection method for selection of a point at random is described in the appendix (p. 84).

A log in the vicinity of P_s is a potential member of the sth sample if there exists a line from the sample point that intersects the central axis of the log perpendicularly. If so, the log is selected into the sample if the perpendicular distance, D, from the central axis of the log to the sample point does not exceed $\kappa_v a$, where a is the cross-sectional area of the log at the point of intersection by the perpendicular line, and κ_v ($\text{m}^2\ \text{m}^{-3}$) is a constant, which is chosen by the investigator (Fig. 6.1).

If the log is curved or branched or both, a straight and horizontal central axis is defined by range poles erected at the farthest end points of the log. Cross-sectional area is measured perpendicular to the central axis and the horizon. If the log is branched, then the cross-sectional area for the log is the sum of the areas for the branches where they are intersected by the perpendicular line. Note, however, that the perpendicular line is most likely perpendicular only to the central axis of the log, and so branch cross-sectional areas are measured on the slant (see the appendix for methods), but perpendicular to the horizon (Fig. 6.2).

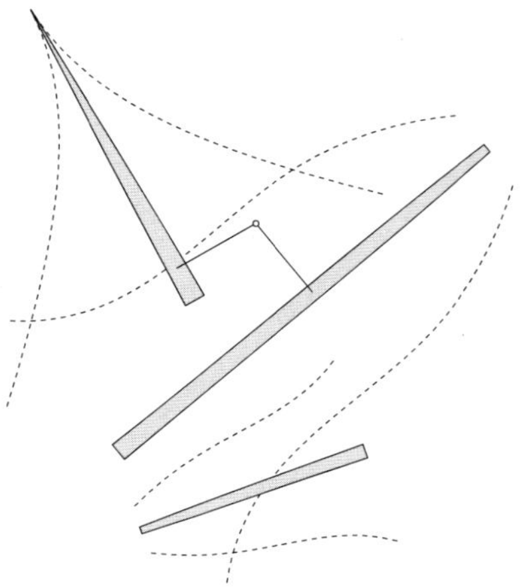

Fig. 6.1 The dashed lines on either side of a log depict the borders of the log's inclusion zone. The area of an inclusion zone is proportional to the log's volume. A log is selected by a sample point (∘) if the sample point occurs in the log's inclusion zone, and this is determined by measuring the perpendicular distance, D, from the log to the sample point. If $D \leq \kappa_v a$, the log is selected. In this example, two logs are selected by the sample point

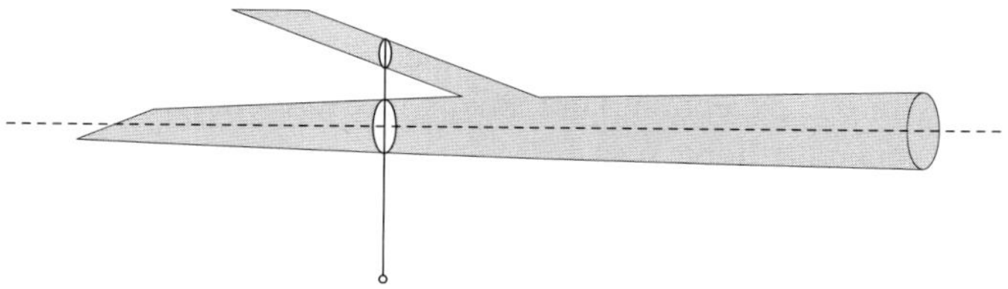

Fig. 6.2 The cross-sectional areas on a multi-stemmed log are measured perpendicular to the central axis of the main stem and coincident with the perpendicular line, and then summed

The probability that a log is included in the sth sample is proportional to the log's volume because cross-sectional area integrated over the length of a log's central axis equals the log's volume, v. Consequently, the probability of inclusion is proportional to $2\kappa_v v$, the factor of 2 arising because the log may be selected by a sample point on either side of the central axis.

The inclusion probability may be diminished if the log is close to a boundary of any region where sample points could not occur, for example, across a property boundary, on a road, in a lake or river, or over the edge of a cliff. Hence, if a selected log, say log k, appears as though it may be closer to a boundary than to the sample point, we suggest implementing the walkthrough method (Ducey et al. 2004, Williams and Gove 2003) to determine whether log k should be tallied twice to correct for the truncation effect by the boundary on the inclusion probability. To wit, measure D_k from the sample point to the perpendicular point on the central axis, then proceed beyond the perpendicular point an additional distance D_k. If a boundary is intercepted before traversing the additional distance, log k is tallied twice, i.e., $t_k = 2$; otherwise, the log is tallied once, i.e., $t_k = 1$.

Let y_k denote an attribute of log k. The aggregate amount of log attribute per unit land area (λ_y) is unbiasedly estimated from the sth sample by

$$\tilde{\lambda}_{y,s} = \frac{1}{2\kappa_v} \sum_{k=1}^{n_s} \frac{y_k t_k}{v_k}, \tag{6.1}$$

where v_k is the volume of the kth log in the sample, and n_s is the number of logs selected into the sample at the sth sample point.

If log volume is the attribute of interest, then $y_k = v_k$, so aggregate log volume per unit land area (λ_v) is unbiasedly estimated from the tallies of the logs in the sth sample, i.e.,

$$\tilde{\lambda}_{v,s} = \frac{1}{2\kappa_v} \sum_{k=1}^{n_s} t_k, \tag{6.2}$$

where $\tilde{\lambda}_{v,s}$ has dimensions cubic meters per square meter. Note that cross-sectional area is not used in the estimator, even though it is measured to determine whether a log is selected into the sample. In the absence of boundary effects, $t_k = 1$ for all k, so

$$\tilde{\lambda}_{vs} = \frac{n_s}{2\kappa_v}. \tag{6.3}$$

Williams and Gove (2003) defined a volume factor $F_v = 10^4/(2\kappa_v)$ ($\text{m}^3\ \text{ha}^{-1}$), which is used to estimate log volume on a per hectare basis. For example, for $\kappa_v = 250\ \text{m}^2\ \text{m}^{-3}$, we obtain

$$F_v = \frac{10{,}000\ \text{m}^2\ \text{ha}^{-1}}{2 \times 250\ \text{m}^2\ \text{m}^{-3}} = 20\ \text{m}^3\ \text{ha}^{-1}.$$

Whence,

$$\hat{\lambda}_{v,s} = 10^4\, \tilde{\lambda}_{v,s} = F_v \sum_{k=1}^{n_s} t_k \tag{6.4}$$

unbiasedly estimates log volume per hectare of land. In the absence of boundary effects,

$$\hat{\lambda}_{v,s} = F_v n_s. \tag{6.5}$$

A combined or 'replicate' estimate obtains from m sample points, i.e.,

$$\hat{\lambda}_{v,\text{rep}} = \frac{1}{m} \sum_{s=1}^{m} \hat{\lambda}_{v,s}. \tag{6.6}$$

The variance of $\hat{\lambda}_{v,\text{rep}}$ is estimated by

$$\hat{v}\left[\hat{\lambda}_{v,\text{rep}}\right] = \frac{\sum_{s=1}^{m} \left(\hat{\lambda}_{v,s} - \hat{\lambda}_{v,\text{rep}}\right)^2}{m\,(m-1)}. \tag{6.7}$$

This estimator is unbiased if the sample points are selected independently at random. The standard error of $\hat{\lambda}_{v,\text{rep}}$ is estimated by

$$\widehat{\text{se}}\left[\hat{\lambda}_{v,\text{rep}}\right] = \sqrt{\hat{v}\left[\hat{\lambda}_{v,\text{rep}}\right]}. \tag{6.8}$$

6.2.1 Example

For illustrative purposes, PDS was applied on a systematic square grid comprising 18 sample points within a 0.8 ha riparian forest fragment in Durham, New Hampshire. The forest succeeded over the past 120 years from a former pasture, and has become dominated by eastern white pine (*Pinus strobus* L.) along with northern red oak (*Quercus rubra* L.), red maple (*Acer rubrum* L.), and eastern hemlock (*Tsuga canadensis* L.).

Logs were tallied at each sample point with a volume factor of $F_v = 14\,\text{m}^3\,\text{ha}^{-1}$ (or, equivalently, 200 $\text{ft}^3\,\text{ac}^{-1}$). This volume factor corresponds to $\kappa_v = 357.1\,\text{m}^2\,\text{m}^{-3}$ (or, equivalently, $\kappa_v = 108.9\,\text{ft}^2\,\text{ft}^{-3}$). A reference table provided the limiting perpendicular distance, D_{limit}, for a log of a given log diameter (d), i.e, $D_{\text{limit}} = \kappa_v \times \pi d^2/4$, where $\pi d^2/4$ approximates log cross-sectional area. Some limiting distances by volume factor and log diameter are provided in Table 6.5 (p. 86).

Since both distance and log diameter are easily ocularly estimated, measurements were taken at a sample point only if a log was borderline, or if the log was forked, because cross-sectional area must be accumulated over the branches. Boundary effects were corrected with the walkthrough method. Logs were categorized by decay class, by diameter class (A, $<$ 14.9 cm; B, 15 – 29.9 cm; C, $\geq$ 30 cm), and type (H, hardwood; S, softwood). Field work for the entire project required approximately 1 h, including walking between points. Tally time per point was typically 2 min or less. Data from the PDS are provided in Table 6.2. Estimates of total log volume per hectare and estimates of hardwood and softwood log volume per hectare, $\hat{\lambda}_{vs}(\text{H})$ and $\hat{\lambda}_{vs}(\text{S})$, respectively, were calcualted with (6.4). The results are provided in Table 6.3. The reader may verify that replicate estimates ($\text{m}^3\,\text{ha}^{-1}$) for the forest fragment are:

$$\hat{\lambda}_{v,\text{rep}} = 30.33, \quad \widehat{\text{se}}\left[\hat{\lambda}_{v,\text{rep}}\right] = 4.56,$$

$$\hat{\lambda}_{v,\text{rep}}(\text{H}) = 3.89, \quad \widehat{\text{se}}\left[\hat{\lambda}_{v,\text{rep}}(\text{H})\right] = 1.52,$$

$$\hat{\lambda}_{v,\text{rep}}(\text{S}) = 26.44, \quad \widehat{\text{se}}\left[\hat{\lambda}_{v,\text{rep}}(\text{S})\right] = 4.92.$$

6.3 Carbon Stock

Of course the real parameter of interest is not volume, but rather the carbon stock per unit land area. Woody dry matter is approximately 50% carbon by weight so, unless bioassay is undertaken, woody biomass is converted to carbon stock with the constant factor $\rho_{\text{C}} = 0.5$ kg C (kg wood)$^{-1}$. The bulk density of wood, i.e., the dry weight of wood per unit volume (kg m^{-3}), may vary among logs because it

Table 6.2 Log data from 18 PDS points in Durham, NH ($F_v = 14$ m^3 ha^{-1})

Point	Species	Decay class	Diameter class	Hollow	Branches	Walk-through	t_k
1	H	3	B	-	-	-	1
2	S	3	C	Yes	-	-	1
3	S	3	B	-	-	-	1
3	S	3	C	Yes	-	-	1
3	S	3	B	-	-	-	1
4	H	4	B	-	Yes	-	1
4	S	4	A	-	-	-	1
5	S	3	B	-	-	-	1
5	S	3	B	-	Yes	-	1
6	S	4	C	-	-	Yes	2
6	H	3	A	-	-	-	1
7	-	-	-	-	-	-	-
8	S	1	C	-	-	Yes	1
8	S	3	B	-	-	-	1
9	S	3	C	-	-	-	1
9	S	4	B	-	-	-	1
10	S	4	B	-	-	-	1
10	S	3	C	-	-	-	1
10	S	4	C	-	-	-	1
11	S	3	B	-	-	-	1
11	S	3	B	-	-	-	1
11	S	3	C	-	-	Yes	2
12	S	4	C	-	-	-	1
12	S	4	A	-	-	-	1
12	S	4	C	-	-	-	1
12	S	3	B	-	-	-	1
13	S	4	B	-	-	-	1
13	S	3	C	-	-	Yes	2
13	S	3	C	-	-	Yes	2
14	S	3	B	-	-	-	1
14	S	3	A	-	-	-	1
15	H	3	B	-	-	Yes	1
16	H	4	B	-	-	-	1
16	S	3	C	-	-	Yes	1
16	S	3	B	-	-	-	1
17	-	-	-	-	-	-	-
18	S	3	B	-	-	-	1

depends on species and degree of decay, among other factors. A core of wood may be extracted from the kth log for the measurement of bulk density ($\rho_{w,k}$), where the volume of the core obtains from the known diameter of the coring device and the 'trimmed length' of the core. Hence the kth log's carbon mass per unit volume (kg C (m^3 wood)$^{-1}$) is estimated by

$$\hat{C}_k = \rho_C \rho_{w,k}. \tag{6.9}$$

Table 6.3 Estimates of log volume, by sample point, calculated from the data in Table 6.2

Point	$\hat{\lambda}_{v,s}$	$\hat{\lambda}_{v,s}$(H)	$\hat{\lambda}_{v,s}$(S)	Point	$\hat{\lambda}_{v,s}$	$\hat{\lambda}_{v,s}$(H)	$\hat{\lambda}_{v,s}$(S)
1	14	14	0	10	42	0	42
2	14	0	14	11	56	0	56
3	42	0	42	12	56	0	56
4	28	14	14	13	70	0	70
5	28	0	28	14	28	0	28
6	42	14	28	15	14	14	0
7	0	0	0	16	42	14	28
8	28	0	28	17	0	0	0
9	28	0	28	18	14	0	14

Our target parameter, carbon stock per unit land area (kg C ha^{-1}), is estimated by

$$\begin{aligned}\hat{\lambda}_{\mathrm{C},s} &= F_v \sum_{k=1}^{n_s} t_k \hat{C}_k \\ &= \rho_{\mathrm{C}} F_v \sum_{k=1}^{n_s} t_k \rho_{w,k}. \end{aligned} \tag{6.10}$$

With m sample points,

$$\hat{\lambda}_{\mathrm{C,rep}} = \frac{1}{m} \sum_{s=1}^{m} \hat{\lambda}_{\mathrm{C},s}. \tag{6.11}$$

The variance of $\hat{\lambda}_{\mathrm{C,rep}}$ is estimated by

$$\hat{v}\left[\hat{\lambda}_{\mathrm{C,rep}}\right] = \frac{\sum_{s=1}^{m}\left(\hat{\lambda}_{\mathrm{C},s} - \hat{\lambda}_{\mathrm{C,rep}}\right)^2}{m\,(m-1)}. \tag{6.12}$$

6.4 Two-phase Sampling

An assumption of homogeneous bulk density among logs may be indistinguishable from wishful thinking in the majority of most, if not all, sampling situations. On the other hand, coring every log may be too much work, or too expensive. And, since coarse woody debris usually is a small carbon pool, compared to the standing trees and soil, it does not deserve a sampling effort or budget that is incommensurate with its importance.

As a compromise, we may estimate the average bulk density, $\bar{\rho}_w$, at a subset of sampling locations or for a subset of logs (perhaps the closest log to the sample point) at all locations, and apply this average to all logs at all locations. Hence, for the sth perpendicular distance sample,

$$\hat{\lambda}_{\mathrm{C,two}} = \rho_{\mathrm{C}} \bar{\rho}_w \hat{\lambda}_{v,\mathrm{rep}} \tag{6.13}$$

estimates the carbon stock in logs per unit land area. The variance of $\hat{\lambda}_{C,two}$ may be estimated with Goodman's (1960) formula

$$\hat{v}\left[\hat{\lambda}_{C,two}\right] = \rho_C^2\left(\bar{\rho}_W\hat{v}\left[\hat{\lambda}_{v,rep}\right] + \hat{\lambda}_{v,rep}\hat{v}\left[\bar{\rho}_W\right] - \hat{v}\left[\hat{\lambda}_{v,rep}\right]\hat{v}\left[\bar{\rho}_W\right]\right). \quad (6.14)$$

The variance of $\bar{\rho}_W$ is problematic, since the inclusion probabilities are proportional to the log volumes, which are unknown. We suggest assuming that the sample cores constitute a random sample from an infinite population, in which case the variance is consistently estimated by

$$\hat{v}\left[\bar{\rho}_W\right] = \frac{\sum_{i=1}^{n}\left(\rho_{W,i} - \bar{\rho}_W\right)^2}{n\left(n-1\right)}, \quad (6.15)$$

where $\rho_{W,i}$ is the bulk density of the ith of n cores in the sample. And, of course, the variance of $\hat{\lambda}_{v,rep}$ is estimated by (6.7).

6.5 Components of Change

The total carbon in coarse woody debris changes over time owing to recruitment, decomposition, and mobility. Recruitment ordinarily results from trees and branches falling to the ground, but in flooded riparian sites, logs may float from off-site to on-site, and vice versa, or change their locations within a tract. The change in the classification of a tree or branch from standing to fallen wood does not, in and of itself, affect the total carbon stock of a tract or the net ecosystem exchange (kg C ha^{-1} year^{-1}). However, the migration of logs to and from the tract and the heterotrophic respiration associated with decomposition do affect both the total carbon stock and the net ecosystem exchange of carbon.

Because a piece of coarse woody debris, by definition, has a minimum size, a log may shrink from coarse to fine woody debris, effecting a loss from the population of logs and an addition to the population of pieces of fine woody debris. Even in the absence of log mobility, transitions from branches to fallen logs to fine woody debris or organic soil complicates the estimation of the contribution of the population of logs to the overall net ecosystem exchange between two points in time.

In the absence of log mobility, the change from time t_1 to t_2 in log carbon per unit land area (Δ_C) is attributable to: (i) the carbon added by recruited logs, which we denote by $\Delta_{C,\text{gain}}$; (ii) the carbon lost from residual logs present in the population at t_1 and t_2, which we denote $\Delta_{C,\text{res}}$; and (iii) the carbon in logs that transition to fine woody debris or organic soil between t_1 and t_2, which we denote by $\Delta_{C,\text{loss}}$. Whence,

$$\begin{aligned}\Delta_C &= \lambda_C(t_2) - \lambda_C(t_1)\\ &= \Delta_{C,\text{gain}} - \Delta_{C,\text{res}} - \Delta_{C,\text{loss}}.\end{aligned}$$

If, during PDS, we measure the distance and azimuth of sample logs at each sample point, P_s, at time t_1 and reuse the same sample point at time t_2, then we can distinguish new logs not present at t_1 from residual logs present in the samples at both t_1 and t_2. Moreover, we can locate and examine each log, which was in the sample at t_1 but absent at t_2, for the purpose of determining whether the log at t_1 still qualifies as a log at t_2, albeit a smaller one, or whether the log at t_1 transitioned to fine woody debris or organic soil by time t_2.

Let $\mathcal{R}_s$ denote the set of residual logs in the samples at P_s at both t_1 and t_2, and let $\mathcal{G}_s$ denote the set of new logs in the sample at t_2. Further, let $\mathcal{L}_s$ denote the set of logs in the sample at t_1, which transitioned to fine woody debris by time t_2, and let $\mathcal{S}_s$ denote the set of logs that were in the sample at t_1 and still classify as logs at t_2, but, owing to shrinkage from decomposition, are not in the sample at t_2. The union of sets $\mathcal{R}_s$ and $\mathcal{G}_s$, i.e., $\mathcal{R}_s \cup \mathcal{G}_s$, comprises all the logs in the sample at t_2, and $\mathcal{R}_s \cup \mathcal{S}_s \cup \mathcal{L}_s$ comprises all the logs in the sample at t_1.

The three components of change are estimated from the samples at P_s with

$$\hat{\Delta}_{\mathrm{C,\,gain},\,s} = \rho_{\mathrm{C}} F_v \textstyle\sum_{k \in \mathcal{G}_s} t_k \rho_{w,k}(t_2) \tag{6.16}$$

$$\hat{\Delta}_{\mathrm{C,\,loss},\,s} = \rho_{\mathrm{C}} F_v \textstyle\sum_{k \in \mathcal{L}_s} t_k \rho_{w,k}(t_1) \tag{6.17}$$

$$\hat{\Delta}_{\mathrm{C,\,res},\,s} = \rho_{\mathrm{C}} F_v \left\{\textstyle\sum_{k \in \mathcal{S}_s} t_k \rho_{w,k}(t_1) + \sum_{k \in \mathcal{R}_s} t_k \left[\rho_{w,k}(t_1) - \rho_{w,k}(t_2)\right]\right\}. \tag{6.18}$$

The investigator may choose to substitute $\bar{\rho}_w(t_1)$ and $\bar{\rho}_w(t_2)$, respectively, for $\rho_{w,k}(t_1)$ and $\rho_{w,k}(t_2)$. And, of course, the total change in the carbon stock per unit land area is estimated by

$$\begin{aligned}\hat{\Delta}_{\mathrm{C},s} &= \hat{\lambda}_{\mathrm{C},s}(t_2) - \hat{\lambda}_{\mathrm{C},s}(t_1)\\ &= \hat{\Delta}_{\mathrm{C,\,gain},\,s} - \hat{\Delta}_{\mathrm{C,\,res},\,s} - \hat{\Delta}_{\mathrm{C,\,loss},\,s}.\end{aligned} \tag{6.19}$$

A combined or replicate estimate of any component of change obtains by averaging the separate estimates from the m sample points, e.g.,

$$\hat{\Delta}_{\mathrm{C,\,loss,\,rep}} = \frac{1}{m} \sum_{s=1}^{m} \hat{\Delta}_{\mathrm{C,\,loss},\,s}. \tag{6.20}$$

The rate of change in the carbon stock in logs per unit land area between t_1 and t_2 is $\Delta_{\mathrm{C}}/(t_2 - t_1)$, which is estimated by

$$\begin{aligned}\frac{\hat{\Delta}_{\mathrm{C,\,rep}}}{t_2 - t_1} &= \frac{\hat{\lambda}_{\mathrm{C,\,rep}}(t_2) - \hat{\lambda}_{\mathrm{C,\,rep}}(t_1)}{t_2 - t_1}\\ &= \frac{\hat{\Delta}_{\mathrm{C,\,gain,\,rep}} - \hat{\Delta}_{\mathrm{C,\,res,\,rep}} - \hat{\Delta}_{\mathrm{C,\,loss,\,rep}}}{t_2 - t_1}.\end{aligned} \tag{6.21}$$

By definition, net ecosystem exchange of carbon (NEE) is the rate of change in the total carbon stock per unit land area (kg C ha^{-1} year^{-1}). Equation (6.21) is an appropriate estimator of the log component of NEE only if this component is defined as the rate of change in the carbon stock in logs per unit land area.

Alternatively, it seems natural to define the log component of NEE as the rate at which carbon from logs is lost to respiration between t_1 and t_2, but owing to the population dynamics of logs, this definition would be problematic. Under the respiration definition, (6.21) is not an appropriate estimator of the log component of NEE. Carbon is added to the log population the instant a large branch or tree falls, but the carbon of interest is not this addition, but rather that which is lost to respiration between the instant of fall and t_2. There is the possibility that a branch may fall and become a log after t_1, but transition to fine woody debris before t_2, in which case the carbon lost between fall and transition is of interest. Residual logs lose carbon between t_1 and t_2, and $\hat{\Delta}_{\mathrm{C,\,res,\,rep}}$ does unbiasedly estimate the change in the carbon stock of residual logs per unit land area. By contrast, $\hat{\Delta}_{\mathrm{C,\,loss,\,rep}}$ subtracts all the carbon in logs that transition to fine woody debris between t_1 and t_2. Of interest, however, is the carbon lost to respiration from each of these logs between t_1 and the time of transition.

6.6 Other Considerations

Carbon stock per unit land area may be estimated by decay class, species, or both. The separate estimates add to provide a total estimate for the tract of interest. However, the use of published estimates of bulk densities by decay class, species, or both, generally would not result in unbiased estimation.

We have suggested PDS as an efficient means to estimate carbon stock per unit land area in coarse woody debris in forested tracts. The superior efficiency of PDS compared to alternative methods, including line intersect sampling, has been demonstrated in coniferous forests in the Western United States, and in northern hardwood forests in New Hampshire. However, these results are not yet published. PDS requires measurements of bulk density from samples of wood, but so do the other methods. However, in its favor, PDS does not require the measurement or estimation of the volumes of the logs in a sample. A measurement of cross-sectional area is needed for selection of a log, but a 'precise measurement' is needed only to determine whether a borderline log is in or out of the sample; the cross-sectional area measurement is not used in the estimator. Log volume per unit area is estimated from the tallies of the logs in the samples.

Since selection of a log into a sample is with probability proportional to volume, samples will comprise, on average, the larger logs in the neighborhood of a sample point. Moreover, since the limiting distance for selection is proportional to log cross-sectional area, the larger sections of these logs are more apt to be intersected by perpendicular lines, and this facilitates coring.

In order to implement PDS, one must decide on a volume factor. A small factor such as 14 m^3 ha^{-1} (or 200 ft^3 ac^{-1}) is feasible only in locations where the logs are small. Doubling the volume factor halves the limiting distance of a given log diameter, so a large volume factor is suggested where large logs are expected (see Table 6.5, p. 86).

Two advances in PDS may be considered as alternatives to the conventional PDS described in this chapter. 'Distance limited PDS' (Ducey et al. 2008a) affords the use of a small volume factor even where large logs occur. In essence, this strategy (a) incorporates a maximum search distance for logs larger than a fixed diameter, (b) requires a diameter measurement on these large logs, and (c) uses a slightly more complicated estimator. Another strategy, called 'omnibus PDS' (Ducey et al. 2008b, Gregoire and Valentine 2008, Chap. 10), provides for efficient estimation of log attributes other than volume, mass, and carbon. Finally, the line intersect distance strategy (LIDS) of Affleck (2008) combines elements of omnibus PDS and line intersect sampling (LIS). Investigators who are already using LIS should find it easy to switch to LIDS.

6.7 Appendix

6.7.1 Acceptance-rejection Method

Surround the region of interest with an imaginary rectangle with dimensions $X \times Z$. Draw two uniform random numbers, u_1 and u_2, with values between 0 and 1. Determine whether coordinates $(u_1 X, u_2 Z)$ occur within the region of interest. If so, accept the location as a sample point or the anchor point of a systematic grid. Otherwise, start over with two new random numbers.

6.7.2 Estimating Cross-Sectional Area on the Slant

Practical methods for estimating cross-sectional area of logs in the field include (cf. Valentine et al. 2001):

(i) Under the assumption that the cross-section of a log is round when viewed perpendicular to its axis of length, the cross-section sliced by an oblique vertical plane will be elliptical unless the plane is parallel to the axis of length or nearly so. Let Ω be the horizontal width of the log sliced (on the slant) by the vertical plane. Let Ω' denote depth of wood calipered vertically in the same plane. Then the cross-sectional area, a, of the log is

$$a \approx \frac{\pi}{4}\Omega\Omega' \tag{6.22}$$

(ii) If Ω' can not be measured, substitute d, which is the diameter calipered horizontally and perpendicular to the axis of length. If the piece of debris is round and not tilted, then d should equal Ω'.

Table 6.4 Symbol table

Symbol	Definition	Units
$a,\ a_k$	Cross-sectional area of log or branch.	m^2
$\hat{C}_k$	Estimated carbon per unit wood volume for the kth log in a sample.	kg C (m^3 wood)$^{-1}$
$d,\ d_k$	Log diameter.	m
t_k	Tally for the kth log in a sample.	count
y_k	Any attribute of the kth log.	
v_k	Volume of the kth log.	m^3
n_s	Number of logs in the sth sample, i.e., the sample at P_s.	count
P_s	The sth of m sample points.	
F_v	Volume factor, which is equivalent to $10^4/(2\kappa_v)$.	$m^3\ ha^{-1}$
$D,\ D_k$	Perpendicular distance from a sample point to the central axis of a log.	m
D_{limit}	Maximum perpendicular distance from a sample point for the inclusion of a log in the sample.	m
Δ_C	Change in log carbon stock per unit area.	kg C ha^{-1}
$\hat{\Delta}_{C,s}$	Estimate of Δ_C from repeated samples at P_s.	kg C ha^{-1}
$\hat{\Delta}_{C,\text{rep}}$	Estimate of Δ_C from m repeated samples.	kg C ha^{-1}
$\Delta_{C,\text{ gain}}$	Gain in log carbon from recruitment.	kg C ha^{-1}
$\hat{\Delta}_{C,\text{ gain},\ s}$	Estimate of $\Delta_{C,\text{ gain}}$ from repeated samples at P_s.	kg C ha^{-1}
$\Delta_{C,\text{ loss}}$	Loss of log carbon by transition to fine wood.	kg C ha^{-1}
$\hat{\Delta}_{C,\text{ loss},\ s}$	Estimate of $\Delta_{C,\text{ loss}}$ from repeated samples at P_s.	kg C ha^{-1}
$\Delta_{C,\text{ res}}$	Loss in log carbon from decomposition.	kg C ha^{-1}
$\hat{\Delta}_{C,\text{ res},\ s}$	Estimate of $\Delta_{C,\text{ res}}$ from repeated samples at P_s.	kg C ha^{-1}
κ_v	Constant design parameter.	$m^2\ m^{-3}$
λ_C	Carbon stock in logs per unit land area.	kg C ha^{-1}
$\hat{\lambda}_{C,s}$	Estimate of λ_C from the sample at P_s.	kg C ha^{-1}
$\hat{\lambda}_{C,\text{rep}}$	Average estimate of λ_C from m samples.	kg C ha^{-1}
$\hat{\lambda}_{C,\text{two}}$	Two-phase estimate of λ_C	kg C ha^{-1}
λ_v	Aggregate log volume per unit land area.	$m^3\ ha^{-1}$
$\tilde{\lambda}_{v,s}$	Estimate of λ_v from the sample at P_s.	$m^3\ m^{-2}$
$\hat{\lambda}_{v,s}$	Estimate of λ_v from the sample at P_s.	$m^3\ ha^{-1}$
$\hat{\lambda}_{v,\text{rep}}$	Average estimate of λ_v from m samples.	$m^3\ ha^{-1}$
ρ_C	Carbon concentration of woody debris.	kg C (kg wood)$^{-1}$
$\rho_{w,k}$	Bulk density of a core from the kth log in a sample.	kg m^{-3}
$\bar{\rho}_w$	Average bulk density in a second-phase sample of cores.	kg m^{-3}
Ω	Horizontal width of a log, coincident with the perpendicular distance line.	m
Ω'	Horizontal width of a log, perpendicular to the log's central axis.	m

Table 6.5 Limiting distance (D_{limit}) by volume factor and log diameter

	F_v (m^3 ha^{-1})				
Diameter	14	28	56	112	224
(cm)			D_{limit} (m)		
2	0.11	0.06	0.03	0.01	0.01
4	0.45	0.22	0.11	0.06	0.03
6	1.01	0.50	0.25	0.13	0.06
8	1.80	0.90	0.45	0.22	0.11
10	2.80	1.40	0.70	0.35	0.18
12	4.04	2.02	1.01	0.50	0.25
14	5.50	2.75	1.37	0.69	0.34
16	7.18	3.59	1.80	0.90	0.45
18	9.09	4.54	2.27	1.14	0.57
20	11.22	5.61	2.80	1.40	0.70
22	13.58	6.79	3.39	1.70	0.85
24	16.16	8.08	4.04	2.02	1.01
26	18.96	9.48	4.74	2.37	1.19
28	21.99	11.00	5.50	2.75	1.37
30	25.24	12.62	6.31	3.16	1.58
32	28.72	14.36	7.18	3.59	1.80
34	32.43	16.21	8.11	4.05	2.03
36	36.35	18.18	9.09	4.54	2.27
38	40.50	20.25	10.13	5.06	2.53
40	44.88	22.44	11.22	5.61	2.80
42	49.48	24.74	12.37	6.19	3.09
44	54.30	27.15	13.58	6.79	3.39
46	59.35	29.68	14.84	7.42	3.71
48	64.63	32.31	16.16	8.08	4.04
50	70.12	35.06	17.53	8.77	4.38
52	75.85	37.92	18.96	9.48	4.74
54	81.79	40.90	20.45	10.22	5.11
56	87.96	43.98	21.99	11.00	5.50
58	94.36	47.18	23.59	11.79	5.90
60	100.98	50.49	25.24	12.62	6.31
62	107.82	53.91	26.96	13.48	6.74
64	114.89	57.45	28.72	14.36	7.18
66	122.19	61.09	30.55	15.27	7.64
68	129.70	64.85	32.43	16.21	8.11
70	137.44	68.72	34.36	17.18	8.59
72	145.41	72.71	36.35	18.18	9.09
74	153.60	76.80	38.40	19.20	9.60
76	162.02	81.01	40.50	20.25	10.13

References

Affleck DLR (2008) A line intersect distance sampling strategy for downed wood inventory. Can J For Res 38:2262–2273

Affleck DLR, Gregoire TG, Valentine HT (2005) Design unbiased estimation in line intersect sampling using segmented transects. Environ Ecol Stat 12(2):139–154

Bebber D, Thomas S (2003) Prism sweeps for coarse woody debris. Can J For Res 33:1737–1743

Brissette JC, Ducey MJ, Gove JH (2003) A field test of point relascope sampling of down coarse woody material in managed stands in the acadian forest. J Torrey Bot Soc 130:79–88

de Vries PG (1986) Sampling Theory for Forest Inventory. Springer-Verlag, Berlin

Ducey MJ, Gove JH, Valentine HT (2004) A walk-through solution to the boundary overlap problem. For Sci 50:427–435

Ducey MJ, Williams MS, Gove JH (2008a) Distance limited perpendicular distance sampling for coarse woody material: theory and field results. Environ Ecol Stat (in press)

Ducey MJ, Williams MS, Gove JH, Valentine HT (2008b) Simultaneous unbiased estimates of multiple downed wood attributes in perpendicular distance sampling. Can J For Res 38:2044–2051

Gove JH, Ringvall A, Ståhl G, Ducey MJ (1999) Point relascope sampling for downed coarse woody debris. Can J For Res 29:1718–1726

Gregoire TG, Valentine HT (2008) Sampling Strategies for Natural Resources and the Environment. Chapman & Hall/CRC Press, Boca Raton, FL

Ståhl G (1998) Transect relascope sampling—a method for the quantification of coarse woody debris. For Sci 44:58–63

Valentine HT, Gove JH, Gregoire TG (2001) Monte Carlo approaches to sampling forested tracts with lines or points. Can J For Res 31:1410–1424

West GB, Brown JH, Enquist BJ (1999) A general model for the structure and allometry of plant vascular systems. Nature 400:664–667

Williams MS, Gove JH (2003) Perpendicular distance sampling: an alternative method for sampling downed coarse woody debris. Can J For Res 33:1564–1579

Williams MS, Valentine HT, Gove JH, Ducey MJ (2005) Additional results for perpendicular distance sampling. Can J For Res 35:961–966

Part III
Measuring Aboveground Carbon Fluxes

Chapter 7
Measuring Litterfall and Branchfall

Pierre Bernier, Paul J. Hanson, and Peter S. Curtis

Abstract Trees capture CO_2 from the air, return a large fraction to the atmosphere via autotrophic respiration, produce live tissues from assimilated carbon, and return expired tissues to the soil via the process of senescence. Methods described in this chapter deal with the measurements of the flow of above-ground solid carbon-based material to the forest floor as litterfall and branchfall. The measurement methods are rather simple and non-technological yet functional. They require an understanding of the scale and dynamics of the processes under study as well as the use to which these data will be put. Back calculation of net primary productivity for example requires not only the mass of foliage captured by litter traps, but also the loss in mass of the leaf between the time it was produced to the time it was captured in the trap. Similarly, the proper capture of ground-area-based large woody debris requires an appropriate spatial sampling design that is tailored to the expected size of the debris to be captured. Stands with a finer branch structure such as fir and spruce will provide a well-distributed rain of twigs and small branches whereas stands of temperate hardwoods will provide small twigs as well as irregular inputs of litterfall in the form of large branches or even tree boles. Alternative methods based on estimation of branch production and retention may also be used to estimate the flux of larger dead material.

Keywords Branchfall, coarse woody debris, fine woody debris, litterfall, net primary productivity

---

P. Bernier
Natural Resources Canada, Canadian Forest Service
PO Box 10380, Stn. Sainte-Foy, Quebec (QC) G1V 4C7 Canada,
E-mail: pbernier@rncan.gc.ca

P.J. Hanson
Environmental Sciences Division
Oak Ridge National Laboratory Bethel Valley Road, Building 1062 Oak Ridge, TN 37831-6244
E-mail: hansonpj@ornl.gov

P.S. Curtis
Department of Evolution, Ecology and Organismal Biology
The Ohio State University 318 W. 12th Avenue Columbus, OH 43210
E-mail: curtis.7@osu.edu

C.M. Hoover (ed.) *Field Measurements for Forest Carbon Monitoring*,

7.1 Introduction

In recent decades, the impact of CO_2 emissions to the atmosphere from anthropogenic sources has permeated many aspects of science. Forestry research is particularly sensitive to this topic because trees are both part of the problem, because of the large CO_2 emission resulting from deforestation activities, and part of the solution because of the capture of atmospheric CO_2 by trees for their growth. It therefore becomes important to develop the capacity to predict not only tree growth in the traditional sense, but also carbon capture and retention by the forest ecosystem as a whole.

Trees capture CO_2 from the air. A large portion of this carbon is sequestered in their long-lived structures, the trunk and the largest branches. The turnaround time for this carbon is low, on the order of many decades to centuries. Traditional forestry has developed ways and means of quantifying the commercially valuable increment in stem volumes or aboveground stem mass. However, more than half of the carbon that is captured as net primary productivity (NPP = gross photosynthesis minus autotrophic respiration) annually may get allocated to more ephemeral structures such as foliage, twigs, branches, bark and non-structural roots (Grier et al. 1981, Schlesinger and Lichter 2001). These components are shed from the trees at either regular intervals, such as hardwood leaves in temperate forests, or irregularly such as the branchfall associated with wind or ice storms (e.g. Ida and Engelmark 2000). Because of its higher surface-to-volume ratio, as well as its higher nitrogen content, foliage and branch litter provides an essential supply of nutrients to the soil pool and is also the most dynamic component of the carbon cycle within the forest ecosystem (Waring and Schlesinger 1985, Schlesinger and Lichter 2001).

In addition to providing information on the flow of material between the different carbon pools, measurements of litterfall and branchfall underpin the back calculation of the NPP of these forest ecosystems. NPP is an important halfway point for modelers between the absorption of carbon by the leaves and the total ecosystem retention of carbon, or net ecosystem productivity (NEP) (e.g. Nemani et al. 2003). NEP can be assessed on the ground through repeated measurements of total ecosystem carbon over an interval of a few to several years, depending on the productivity of the site. However, the modeling of net ecosystem carbon exchange from first principles is a complex process that involves a cascade of modeling steps starting with gross photosynthesis, respiration, mortality of live biomass plant components, and an estimation of within stand tree mortality, or self thinning. Accumulation of errors through the modeling process makes direct comparison of modeled NEP with field measurements difficult. Field experimentalists can backtrack from NEP to NPP by adding back to the accrual in plot-level biomass that is usually obtained through allometric equations, the mass of plant parts that have been shed over the course of a given time period, generally one year. These assessments provide key values for modelers to verify the accuracy of their approach at an intermediate complexity level.

The methods described in this chapter deal with the measurements of the flow of above-ground solid material to the forest floor as litterfall and branchfall, as well as the decomposition of these plant parts and their release of carbon back to the atmosphere and of nutrients to the ecosystem. The methodologies of the measurements used to characterize these processes are rather simple and non-technological, but require an understanding of the scale and dynamics of the processes under study as well as the use to which these data will be put. Back calculation of NPP, for example, requires not only the mass of foliage captured by litter traps, but also the loss in mass of the leaf between the time it was produced to the time it was captured in the trap. Similarly, the proper capture of larger debris requires an appropriate spatial sampling design that is tailored to the expected size of the debris to be captured. Stands with a finer branch structure such as fir and spruce will provide a well-distributed rain of twigs and small branches whereas stands of temperate hardwoods will provide small twigs as well as irregular bursts of material in large branches.

Finally, patterns of decomposition provide an essential link between the aboveground components and both nutrient cycling and release of the carbon back to the atmosphere. This last link is particularly important in light of the importance of soil and litter carbon stocks in boreal and temperate forests, and of the uncertainty concerning the impact of climate change on decomposition processes (Moore et al. 1999, Price et al. 1999). Measurements of rates of fine and coarse litter decomposition, as described in Chapter 8, will provide key information to help constrain models and provide better insights into future forest contribution to climate change.

7.2 Quantifying Litterfall

The annual production of canopy foliage and its subsequent senescence, coupled with terminal branch, flower, and fruit/seed production and senescence, represent a key annual sink for carbon in forest and grassland ecosystems. Mass data from collected litter can provide direct information about annual carbon inputs to the forest floor. For annual grasslands and deciduous forests these data provide a direct estimate of canopy production within a given annual cycle. For coniferous systems annual cumulative litter collections can include annual production from more than one annual cycle, but long-term average annual litter inputs to coniferous systems should be related to annual canopy production.

Leaf litter mass data are also the primary set of measurements from which maximum annual leaf area index can be derived when corrected for losses from herbivory. While indirect measurement methods for leaf area display are available (Chason et al. 1991, Breda 2003), they do not provide a direct measure of canopy mass and are therefore not recommended as a measure of canopy production.

This protocol addresses the determination of mass of each of the following components of canopy production: leaves/needles, bracts and flowers, seeds/fruits, and small branches. Annual pollen production and the production of volatile organic carbon emissions (VOCs such as isoprene) are minor component of annual canopy production and could be evaluated (Kesselmeier et al. 2002, Pressley et al. 2005,

Funk et al. 2006), but they will be ignored at landscape-scale monitoring sites as a logical cost-saving tradeoff.

7.2.1 Sampling Design

Litter is often collected in baskets or mesh containment devices placed just above the ground. Litter collection baskets (Fig. 7.1) or trays must be randomly positioned within subplots to avoid arbitrary influences of individual tree canopies. Baskets might be placed a fixed distance from all subplot centers at a randomly chosen or fixed compass direction.

The appropriate level of replication and sampling design will vary with vegetation type, but any given design must cover the range of stand conditions. Replicate litter collection baskets at each landscape-scale monitoring location should be sufficient in number to allow detection of a 20% difference in the inter-annual litter fall at a given site. For a typical deciduous hardwood forest this would require approximately 15–20 litter baskets per location. The proposed landscape-scale monitoring site layout with 16 inventory plots for each monitoring location will accommodate this requirement (16 baskets per location – one per plot).

The timing of litter collections throughout the year will depend on the vegetation type being studied. Deciduous forest collections in the northern hemisphere need to be made in late spring and throughout the autumnal senescence period. Conifer litterfall collections need to be conducted at appropriate times throughout the year

a b

Fig. 7.1 Litter collection basket appropriate for large hardwood leaves (left) and a litter mesh-based collection device appropriate for litterfall of smaller diameters (right) to minimize degradation of litter materials between collections. During periods of maximum litter inputs (i.e., fall senescence) litter collections should be done on a minimum of a biweekly interval to avoid prolonged exposure to rainfall and the resulting decomposition of standing litter. Since the greatest laboratory costs are associated with sorting, and are therefore a direct function of the volume of the material to be treated, more collections per year should not overly increase cost.

Commercially available laundry baskets can be a cost effective collection device for hardwood leaf litter. Holes need to be drilled into the bottom of these baskets to allow drainage of precipitation, and fine mesh netting may need to be added to avoid losing small plant parts or needles in conifer ecosystems. A typical rectangular basket (round baskets are also acceptable) having collection area of 53×38 cm produces a total collection area of 0.2 m^2 and has proven very effective (Hanson et al. 2003). All baskets must be positioned in a permanent location for the duration of the landscape-scale monitoring observations, and should be elevated slightly off the ground surface. Inexpensive lengths of small diameter PVC pipe (~1 cm) can be cut for this purpose. Three legs per basket or tray serve to mark the permanent location of the collection device, elevate it off the ground, and keep the basket from blowing over in moderate winds.

At designated collection intervals, all materials from each basket should be collected in paper bags labeled with the date of collection and a site and subplot identification name/number. These materials should be dried (see below) as soon as possible after their collection in the field.

7.2.2 Sampling Problems

The primary problem with litter basket collections is the potential loss of samples from disturbance of the collection device. This problem can be minimized if the baskets are checked prior to major collection events. For example, prior to fall senescence the condition of all litter collection baskets should be checked.

A second problem is the diameter cutoff for twig collection. Most protocols use 1 cm as the standard. Small twigs with diameter at their big end <1 cm and that fall totally within the basket cause no problem. However, twigs still attached to branches that extend beyond the basket should be rejected, and that component should be measured with the branchfall traps (see below). This procedure is adopted because branches will tend to bounce off the basket sides, causing an underestimation of small twigs in the baskets. It is therefore better to capture larger branches and their associated small twigs in the branchfall traps.

7.2.3 Sample Processing

Once collected, the litter samples should be returned to the laboratory to be dried at 70°C for 24–48 h or until mass loss is complete. Individual litter samples should be sorted into leaves, twigs/branches <1 cm diameter, woody material, seeds, and flowers in all years. To facilitate the calculation of maximum leaf area and the contribution of species to canopy leaf production, leaf litter should be sorted by species for at least one annual cycle, but possible inter-annual variations could justify this process more often. Sorting is easier and results in a lower miscellaneous species component (from leaf fragmentation) if done before oven drying, but it is often not feasible to accomplish this time-consuming task quickly enough following collections. Successful species separations can be accomplished on oven-dried or air-dried material.

The area and mass of leaves from the main tree species within the plot (to be determined from a summary of stand basal area contribution) should be obtained on a sub-sample of litter in order to calculate their average litter mass per unit leaf area (ρ_L or LMA in gram per square centimeter). These data will be used to estimate maximum canopy leaf area index, and also to calculate seasonal patterns of LAI (Chapter 14 this volume). Bulk leaf, twig, flower/bract, and seed pools should be analyzed for [C] and [N] to calculate litter content of these elements if funding allows. A typical value for leaf [C] is just under 50%. This information will be essential for converting litter mass inputs to C production numbers.

7.2.4 Data Processing

Canopy production should be reported in units of grams Carbon per square meter per year by component (i.e., leaves, twigs/branches, flowers/bracts, seeds). The annual production of canopy components (P_c, g m^{-2} year^{-1}) can be determined from the following equation:

$$P_c = \frac{M_c\,(1-H)\,C_c\left(\dfrac{\rho_{L,f}}{\rho_{L,l}}\right)}{A} \tag{7.1}$$

where M_c is the annual dry mass of the litter component (g a^{-1}), H is the fraction of leaf area loss to herbivory, C_c is the carbon fraction of dry mass for the component of interest, $\rho_{L,f}$ and $\rho_{L,l}$ are the average leaf mass per unit area of fresh leaves (a canopy average) and of leaves in the litter respectively, and A is the total basket area (m^2). The LMA correction for the difference between green-leaf and litterfall mass per unit area is needed because there is a substantial mass loss during leaf senescence due to translocation of leaf resources (e.g., amino acids) from leaves to storage locations within plant branches and stems. Results on balsam fir and sugar maple suggest a mass loss of 20–30%.

For calculating inputs into the soil compartment, Eq. 7.1 is simplified to:

$$P_c = \frac{M_c\,C_c}{A} \tag{7.2}$$

Seasonally maximum leaf area index (LAI) of deciduous vegetation can be obtained from a simple multiplication of the species-weighted LMA (g m^{-2} leaf area) times the cumulative mass of annual leaf production (g m^{-2} ground area) as obtained from Eq. 7.1. An example of the species weighted mean LMA calculation and the calculation of maximum LAI is as follows, assuming that the total leaf mass collected in the litter traps was 500 g m^{-2}:

Species	*Fraction of Leaf litter mass*	*Species-specific-LMA (g m^{-2}leaf-litter area)*	*LMA contributed by a species*
Quercus	0.51	93	= 47.4
Acer	0.25	63	= 15.7
Misc.	0.24	80	= 19.2
Mean species-weighted LMA = (sum of species values)			= 82 *g m^{-2}leaf-litter area*

$$LAI_{\max} = \frac{Total\ leaf\ litter\ mass}{Mean\ species\text{-}weighted\ LMA} = \frac{500 g\, m^{-2}}{82 g\, m^{-2}} = 6.1 m^2\, m^{-2}$$

This calculation assumes limited loss of area between green leaves and leaf litter if re-hydrated.

7.2.5 Reporting the Data

The data should be reported as yearly averages of canopy production of leaves (Eq. 7.1), of leaf litter input (Eq. 7.2), and of litter capture of flowers, fruits/seeds, and branches < 1 cm diameter, all expressed in grams per square meter of dry mass and of carbon. Periodic measurements of individual baskets and from individual dates should be kept as permanent records. Other supporting measurements, not covered by the regular forest inventory include the canopy-average leaf mass per unit area per species on the plot for use in Eq. 7.1, and the estimation of herbivory loss. Phenological studies using above- and below-canopy light capture (e.g. Raulier and Bernier 2000) would also greatly complement these measurements.

7.3 Quantifying Branchfall

We define branchfall (*B*) as the loss of woody material of any diameter class that is not associated with whole-tree mortality and that is not properly included as a component of annual fine litterfall (see above). Branchfall occurs both as a result of normal tree ontogeny and due to episodic disturbance. It is of interest in forest carbon flux estimates for two reasons. First, branchfall must be quantified to properly account for all inputs to litter and soil carbon pools. Secondly, branchfall must be used to obtain appropriate estimates of carbon allocation to branch growth for field estimates of net primary productivity. The latter is the case when the annual increment in woody biomass (M_A) is estimated as the difference in individual tree woody mass (M_T) between two census intervals and when M_T is estimated from allometric equations derived from harvested trees that also have lost mass as *B*.

$$M_{T2} = M_{T1} + M_A - B \quad (7.3)$$

The addition of *B* is necessary since trees harvested at random for allometric analysis would be expected to have experienced such mass loss after the production of these structures. However, caution is needed for estimates of NPP for a given year since branchfall amounts may be related to unusual disturbance events, and hence only long-term average of branchfall may yield proper estimates of annual *B*.

How important this correction is depends, of course, on the magnitudes of *B*, M_A, and other sources of error or uncertainty. For a tropical moist forest, Clark et al. (2001a) calculated a potential impact on estimated NPP of between 0% and –17% by not correcting for *B* and heartrot (another source of mass loss), compared to between +11% and –20% for using inappropriate mass allometry. Woody material as a component of annual litterfall has been reported as averaging 30% in boreal conifers and 20% in *Populus tremuloides* (Bernier et al. 2007), 20% (range 4–33%) in tropical forests (Clark et al. 2001b), 18% in a eucalyptus forest and 6% in a young *Pinus radiata* plantation (Putuhena and Cordery 1996), 21% (13–33%) in a mixed northern hardwood forest (Gosz et al. 1972), and 45% in an old-growth *Pseudotsuga menziesii* forest (Grier and Logan 1977). Clearly, sites should be evaluated on a case-by-case basis in deciding on the need for *B* corrections to estimated M_A. An alternative method for estimating *B* from that of directly sampling downed branches, as described below, was suggested by Bernier et al. (2007). They estimated live branch mass from published allometric equations and annual branch NPP from measured annual foliage NPP. Branchfall between two measurement dates was then equal to branch NPP during that period minus the live branch mass increment. This method requires robust empirical relationships between branch NPP and foliage NPP, which may be difficult to obtain, but avoids the sometimes significant sampling problems associated with spatial and temporal heterogeneity in *B*.

7.3.1 Sampling Design

Depending on the diameter size class distribution of *B*, a combination of fine litterfall traps, forest floor branchfall traps, and line transect or permanent plot resurvey methods for coarse woody debris assessment may be required. Litterfall traps are considered elsewhere and generally are adequate for assessing $B \leq 1$ cm diameter. For debris with butt-end diameters above 5 cm, we recommend using periodic line intersect sampling techniques. The low frequency of large branchfall makes them difficult to capture with a sparse sampling design.

Branchfall traps for material >1 cm diameter (or the upper size limit of litterfall traps) and <5 cm diameter (or the lower limit of CWD assessments) consist of 2×2 m squares of landscape cloth pinned to a previously cleared area of ground (Fluxnet-Canada 2003). An alternate design is to permanently mark the corners of a cleared, 2×2 m sampling quadrat. The exact number of traps required will depend on the spatial variability in *B*. Gosz et al. (1972) found that 15 2×2 m traps within a 13.2 ha watershed were insufficient to sample branchfall <10 cm diameter within a standard deviation of 10%. A minimum of three traps per landscape-scale monitoring plot (75 traps ha^{-1} for 400 m^2 plots) is recommended.

Visits to the traps should be made annually. In snow-dominated regions, an early spring visit would permit the capture of winter branchfall. Visits should also be made following unusually violent storm events. Traps need to be recleared of all woody debris following measurement.

7.3.2 Sampling Problems

If the 1 cm diameter rule was perfect for partitioning between litter and branchfall collection, one should remove from each branch all portions with a diameter below 1 cm. However, because of problems with branches bouncing off the sides of the litter baskets we recommend that all components of branches be kept and processed. Only small branches whose butt-end is less than 1 cm should be rejected, as these are assumed to be well captured in the litter baskets. Local variation of these rules may be necessary, the important thing being the proper accounting of all litter types and size fractions.

Branches crossing the cloth edge should be cut off at the edge, and only the part within the trap considered. With branches with a butt-end >5 cm, all parts with a diameter <5 cm and enclosed within the trap should be trimmed and collected. Parts with diameter >5 cm should be accounted for in yearly sampling along line transects on which all material is cleared at each visit. The protocol for coarse woody debris (Chapter 6) may be followed for this purpose.

Traps should be visited periodically to make sure that the corners of the landscape cloth are still well pinned to the ground, or the corner markers if bare ground is used are still well visible.

7.3.3 Sample Processing

Recovered material is bagged, air dried to constant mass, and weighed. Grinding of a sub-sample and determination of C and N concentrations is strongly suggested if budgets permit.

7.3.4 Data Processing and Reporting

Computation of B in gC m^{-2} yr^{-1} is performed as:

$$B = \frac{B_{dm}\,C_c}{A} \tag{7.4}$$

where B_{dm} is the total dry mass of branches collected in all traps on a given year, C_c is the carbon concentration in the samples, and A is the total trap area. These numbers should be reported along with litter mass.

Literature Cited

Bernier PY, Lavigne MB, Hogg EH, Trofymow JA (2007) Estimating branch production in trembling aspen, Douglas-fir, jack pine, black spruce, and balsam fir. Canadian Journal of Forest Research 37: 1024–1033

Breda NJJ (2003) Ground-based measurements of leaf area index: a review of methods, instruments and current controversies. Journal of Experimental Botany 54: 2403–2417

Chason JW, Baldocchi DD, Huston MA (1991) A comparison of direct and indirect methods for estimating forest canopy leaf area. Agricultural and Forest Meteorology 57: 107–128

Clark DA, Brown S, Kicklighter DW, Chambers JQ, Thomlinson JR, Ni J (2001a) Measuring net primary production in forests: concepts and field methods. Ecological Applications 11: 356–370

Clark DA, Brown S, Kicklighter DW, Chambers JQ, Thomlinson JR, Ni J, Holland EA (2001b) Net primary production in tropical forests: an evaluation and synthesis of existing field data. Ecological Applications 11: 371–384

Fluxnet-Canada (2003) Fluxnet-Canada measurement protocols. Working draft version 1.3. Fluxnet-Canada Network Management Office, Université Laval, Québec, Canada

Funk J, Giardina C, Lerdau M (2006) The influence of nutrient availability and stand age on leaf- and canopy-level isoprene flux in a *Eucalyptus saligna* experimental forest. Journal of Geophysical Research 111, G02012, doi:10.1029/2005JG000085

Gosz JR, Likens GE, Bormann FH (1972) Nutrient content of litter fall on the Hubbard Brook Experimental Forest, New Hampshire. Ecology 53(5): 769–784

Grier CC, Logan RS (1977) Old-growth *Pseudotsuga menziesii* communities of a western Oregon watershed: biomass distribution and production budgets. Ecological Monographs 47: 373–400

Grier CC, Vogt KA, Keyes MR, Edmonds RL (1981) Biomass distribution and above- and below-ground production in young and mature *Abies amabilis* zone ecosystems of the Washington Cascades. Canadian Journal of Forest Research 11: 155–167

Hanson PJ, Todd DE, Joslin JD (2003) Canopy production. In: Hanson PJ, Wullschleger SD (Eds.) North American temperate deciduous forest responses to changing precipitation regimes. Springer, New York, Ecological Studies 166: 303–315

Hanson PJ, Edwards NT, Tschaplinski TJ, Wullschleger SD, Joslin JD (2003) Estimating the net primary and net ecosystem production of a southeastern upland *Quercus* forest from an 8-year biometric record. In: Hanson PJ, Wullschleger SD (Eds.) North American temperate deciduous forest responses to changing precipitation regimes. Springer, New York, Ecological Studies 166: 378–395

Ida H, Engelmark O (2000) Treefall gap disturbance in an old-growth beech forest in southwestern Japan by a catastrophic typhoon. Journal of Vegetation Science 11(6): 825–832

Kesselmeier J, Ciccioli P, Kuhn U, Stefani P, Biesenthal T, Rottenberger S, Wolf A, Vitullo M, Valentini R, Nobre A, Kabat P, Andreae MO (2002) Volatile organic compound emissions in relation to plant carbon fixation and the terrestrial carbon budget. Global Biogeochemical Cycles 16: 1126–1134

Moore TR, Trofymow JA, Taylor B, Prescott C, Camire C, Duschene L, Fyles J, Kozak L, Kranabetter M, Morrison I, Siltanen M, Smith S, Titus B, Visser S, Wein R, Zoltai S (1999) Litter decomposition rates in Canadian forests. Global Change Biology 5: 75–82

Nemani RR, Keeling CD, Hashimoto H, Jolly WM, Piper SC, Tucker CJ, Myneni RB, Running SW (2003) Climate-driven increases in global terrestrial net primary production from 1982 to 1999. Science 300: 1560–1563

Pressley S, Lamb B, Westberg H, Flaherty J, Chen J, Vogel C (2005) Long-term isoprene measurements above a northern hardwood forest. Journal of Geophysical Research 110:D07301, doi:07310.01029/02004JD005523

Price DT, Peng CH, Apps MJ, Halliwell DH (1999) Simulating effects of climate change on boreal ecosystem carbon pools in central Canada. Journal of Biogeography 26: 1237–1248

Putuhena WM, Cordery I (1996) Estimation of interception capacity of the forest floor. Journal of Hydrology 180: 283–299

Raulier F, Bernier PY (2000) Predicting the date of leaf emergence for sugar maple (*Acer saccharum* Marsh) across its native range. Canadian Journal of Forest Research 30: 1429–1435

Schlesinger WH, Lichter J (2001) Limited carbon storage in soil and litter of experimental forest plots under increased atmospheric CO_2. Nature 411: 466–469

Waring RH, Schlesinger WH (1985) Forest ecosystems: Concepts and Management. Academic Press Inc, Orlando, 340 p

Chapter 8
Methods for Estimating Litter Decomposition

Noah J. Karberg, Neal A. Scott, and Christian P. Giardina

Abstract Litterfall in terrestrial ecosystems represents the primary pathway for nutrient return to soil. Heterotrophic metabolism, facilitated through comminution by small insects and leaching during precipitation events, results in the release of plant litter carbon as CO_2 into the atmosphere. The balance between litter inputs and heterotrophic litter decomposition influences the amount of carbon stored in the forest floor. Periodic measurements of litterfall and litter decomposition with standard techniques will provide much needed information on carbon and nutrient cycling in forests. These available methods include mass balance, litterbags, tethered leaves, and the cohort layered screen. One must consider the strengths and limitations of each method as applicable to the goals of the study, and apply the most appropriate method, or combination thereof. For all methods, sufficient replication is required to accurately estimate stand level decomposition, and site selection for deployment should represent the various microsites likely to be encountered in the forest stand being examined.

Keywords Cohort screen, litterbags, litterfall, mass balance, tethered leaves

N.J. Karberg
US Forest Service, Northern Research Station, 410 MacInnes Drive, Houghton, MI 49931
E-mail: nkarberg@fs.fed.us

N.A. Scott
Woods Hole Research Center, Falmouth, MA, Current Address: Department of Geography, Queen's University, D201 Mackintosh-Corry Hall, Kingston, ON K7L 3N6, Canada
E-mail: scottn@queensu.ca

C.P. Giardina
US Forest Service Pacific Southwest Research Station, 60 Nowelo Street, Hilo, HI 96720
E-mail: cgiardina@fs.fed.us

C.M. Hoover (ed.) *Field Measurements for Forest Carbon Monitoring*,

8.1 Introduction

In terrestrial systems, plant litterfall is a primary pathway for the return of nutrients to the soil. Leaf tissue can account for 70% or more of aboveground litterfall in forests, with the remainder composed of stems, small twigs and reproductive structures (Robertson and Paul 1999). Litter decomposition proceeds by several mechanisms including heterotrophic utilization of organic compounds in litter, but also leaching during rain events and comminution by small insects which do not lead directly to CO_2 release to the atmosphere. The release of plant litter carbon (C) as CO_2 through heterotrophic decomposition by soil microorganisms can contribute 20% or more to soil surface CO_2 efflux, which is often referred to as soil respiration (Chapter 11, this volume, Raich and Nadelhoffer 1989). The balance between litter inputs and heterotrophic litter decomposition influences the amount of C stored in the forest floor; this is important because forest floor C can respond to disturbance over short time scales (e.g. Gaudinski et al. 2000). Further, nitrogen (N), phosphorus (P), and calcium (Ca) are released from plant litter during decomposition where they can become available for plant and microbial uptake. Given the important role of litter decomposition to C storage and tree nutrition, it is important to quantify litter decomposition rates for accurate characterization of forest carbon dynamics.

Litter decomposition rates are controlled by three main factors: temperature, moisture, and litter quality. Faunal community structure, especially the influence of earthworms, is increasingly being recognized as a possible fourth important factor (Bohlen et al. 1997, Dechaine et al. 2005). Where substrate is available, soil microbial activity increases exponentially with soil temperature, with microbial activity often doubling with a 10°C increase in temperature (Kirschbaum 1995). Microorganisms can also be limited by soil moisture. As temperatures increase, soil moisture assumes an increasingly important role for maintaining high rates of microbial activity (Peterjohn et al. 1994). As a result, rates of fresh litter decomposition increase with both increasing temperature and precipitation (Meentemeyer 1978).

This general pattern of decomposition can also be influenced by variability in litter quality. Quality refers to characteristics of the litter (chemistry, physical attributes, etc.) that influence the susceptibility of litter to decomposition. Litter containing high concentrations of labile compounds (e.g. sugars, amino acids) tends to decompose rapidly because these compounds can be readily metabolized by soil microorganisms or leached. For example, labile structural compounds such as cellulose are quickly cleaved by exoenzymes into sugar sub-units, which again are readily metabolized by microbial organisms. In contrast, recalcitrant structural compounds such as lignin and chitin are too large to pass through cell membranes, and are instead slowly processed by extracellular enzymes. Irregular chemical structure and complicated bonding make these compounds difficult for enzymes to attack, providing a slow release of N and P for continued microbial growth.

Three hypotheses are proposed to explain how initial litter quality influences litter decomposition and N release from decomposing litter (references as cited in Giardina et al. 2001). The first hypothesis suggests that litter decomposition and N release are positively related to initial litter quality. In the early stages of decomposition the ratio of C:N may be the best predictor of mass loss and N release, with lignin content becoming increasingly important at later stages of litter decomposition. In the decay filter hypothesis, differences in initial litter quality (such as the ratio of lignin:N and lignin:cellulose) alter litter decomposition and release rates in the early stages of litter decomposition. As litter substrate quality decreases during decomposition, initial litter quality has a decreasing influence on late-stage decomposition rates. At this stage, litter decay rates are controlled instead by climate, soil texture, and exogenous sources of labile C and nutrients. The third hypothesis suggests that litter decomposition and rates of N release are negatively related to N-based estimates of initial litter quality. High N content may actually retard litter decomposition rates later in the decomposition process, particularly if lignin levels are also high. Regardless of the underlying mechanisms, periodic measurements of litterfall and litter decomposition with standard techniques will provide much needed information on C and nutrient cycling in forests.

8.2 Available Methods

8.2.1 Mass Balance

Mass balance techniques are used to estimate litter decomposition for whole ecosystems, and are often employed when direct measurement is too cumbersome or expensive. When applied to aboveground litter decomposition, the mass balance approach suggests that annual litter decomposition should equal the annual input of fresh litter as long as the mass of detrital litter stored in the ecosystem remains constant (Olsen 1963, Schlesinger 1997). This approach assumes that a constant fraction, *k*, of the detrital litter mass decomposes, where

$$\text{litterfall} = k\ (\text{detrital litter mass}),\ \text{or}$$
$$\text{litterfall/detrital litter mass} = k.$$

For example, if the mass of the forest floor is 10 Mg C ha^{-1} and annual litterfall is 1.0 Mg C ha^{-1} year^{-1}, then litter decay rate would equal 0.1 year^{-1}. In forest ecosystems where decomposition rates are rapid and there is little surface litter accumulation, values for *k* are greater than 1.0. Ecosystems with slow decomposition rates and surface litter accumulation, for comparison, have *k* values that are less than 1.0.

From the equation, the method requires the collection of two variables: litterfall and detrital litter mass. Litterfall is measured using litter traps that are randomly spaced as appropriate throughout the study site (see Chapter 7, this volume, Bubb et al. 1998, Xu and Hirata 2002). Typically, litter trap openings are from 0.5 to 1 m across, and litter is contained within the trap by netting or a mesh screen. Traps are emptied at biweekly to monthly intervals, and collected materials can be sorted into categories by litter type, species, and/or component, oven dried, and then weighed. Detrital litter mass, often called the forest floor and defined by USDA Soil Survey soil taxonomy as the Oi, Oe and Oa horizons, is estimated by removing the forest floor from a known area and, after drying the material, determining the dry weight. These quadrats are typically 1 x 1 m^2, though their dimension and frequency of measurement should be determined by the study site and objectives of the study. The forest floor is collected from inside the quadrat and sorted by component. The entire sample can be oven dried for dry weight determination, or wet weights can be measured and a subsample taken for dry weight determination. Because the Oa can contain up to 20% mineral mass, and soil can contaminate upper forest floor layers, forest floor mass should be corrected by sample combustion to determine the ash-free portion of the sample.

The mass balance approach can be used independently to estimate litter decomposition, or as a check on model predictions (Hedin 2000). It provides a robust estimate of litter decomposition at the stand level, though assumptions about steady-state stand conditions and constant forest floor decomposition dynamics complicates interpretation of calculated litter decay rates. Mass balance based estimates of litter decomposition are imprecise where short-term (e.g., annual) estimates are needed but forest floor mass is not in steady state. This method may not be appropriate in young stands where the forest floor is rapidly aggrading. In this case, the method would over-estimate decomposition rates. Because the method relies on native litterfall, this approach cannot be used to cleanly elucidate the role of other factors such as temperature and moisture as can common-litter litterbag experiments.

8.2.2 Litterbags

The litterbag approach is widely used to study decomposition at the soil surface. Fresh leaf litter is enclosed in mesh bags, placed on the ground, and collected at periodic intervals for measurement of the mass remaining. A subset of the collected litter is oven dried to later establish wet to dry conversions for comparison. Mesh size is generally chosen to optimize access by all organisms to the litter while minimizing excessive particle loss, though mesh size can also be manipulated to exclude functional groups of litter decomposers. Very small mesh size will not only exclude certain organisms, but hinder particle loss to mineral soil as well. Fiberglass mesh has been recommended for light intensive sites where UV light will degrade nylon and other materials (Harmon and Lajtha 1999). Though 1–2 mm mesh is most common in litterbag studies (Robertson and Paul 1999), litterbag mesh size should be greater than 2 mm if a goal is to allow entry by macrofauna. Specific procedures for assessing the contributions of macroinvertebrates to decomposition can be found in Coleman et al. (1999).

Size and content of the litterbags is also an important component of litterbag studies. Overall bag size should be appropriate to the litter-specific ecosystem under consideration. While 20 x 20 cm bags are common (Robertson and Paul

1999), diverse plant communities or large leaf sizes may call for a larger litterbag. Litter should be freshly senesced. Litterbags are typically constructed with only one species, but when a more realistic experiment is desired, litterbags can be made with a proportionally representative mix of species litter and even small woody debris or reproductive structures.

The number of litterbags deployed at a site will depend on the variability of the site, the number of collections per year, and the number of years of the study. A forest with heterogeneous microclimate and stand characteristics will require a greater number of litterbags than an even-aged plantation to accurately calculate *k*. Typically, five or more replicate litterbags are collected at each sampling interval during the first year of the study, with two to four collections in subsequent years. This allows for a more robust characterization of the decay curve. Again, variability in stand micro-environment and overstory/understory diversity and associated litter quality should be considered before decisions about the number of required replicates is made.

Collected litterbags are oven dried in order to compare pre-and post-decomposition sample mass: separate samples may need to be freeze dried if substrate-specific chemistry will be analyzed. As with forest floor samples, mineral soil often contaminates litterbag samples, and should be corrected for by measuring the ash content of litter before and during decomposition. Litter decomposition rates are often estimated using a regression approach and the first order negative exponential decay equation, where the fraction of litter remaining after 1 year is given by:

$$X_t / X_o = e^{-kt} \tag{8.1}$$

where X_t/X_o is the proportion of original mass remaining at time *t*, and *k* is the decomposition rate constant. The decomposition rate constant, *k*, can be calculated by fitting the exponential decay model to a scatter plot of *t* vs. X_t/X_o (e.g. Harmon et al. 1999).

An alternate modified double exponential model that can provide a better fit for decomposition over the long term is given by:

$$Y = 100 \exp(-kt^p) + \varepsilon \tag{8.2}$$

where Y is the original mass remaining at time *t*, *k* is the decomposition rate constant, *p* is a parameter allowing the mass loss rate to change with time, and ε is the random component with a mean of 0 and variance of σ^2 (Kelly and Beauchamp 1987, Hanson et al. 2005). The decomposition rate constant *k* and parameter *p* are estimated from the data.

Litterbags have a few weaknesses and caveats regarding their use. Certain macroinvertebrates are excluded from the litterbags, lowering rates of litter comminution. Contamination by soils with high organic matter contents requires corrections. Care needs to be taken to ensure that the litterbag represents a realistic mixture of litter species and components, and that bag placement does not alter the microclimate or decomposition conditions. Despite these limitations, litterbags represent a classic approach to estimating decomposition rates in the field, in particular because they can be used experimentally to quantify rates at various time scales and the contribution of different factors (e.g. temperature, moisture content). There have been many published litterbag studies, providing a rich database for comparison of results (see Vitousek et al. 1994).

8.2.3 Tethered Leaves

The tethered leaf approach is similar to the litterbag approach, except that individual leaves are tied together in bundles rather than placed in litterbags. Either a single leaf, or groups of leaves, are tied together using nylon thread or monofilament fishing line. The line is tied to the leaf petiole for durability: the line is usually anchored to both a reference point for collection, and an identifying tag.

A "wheel spoke" approach modeled after Vitousek et al. (1994) is often employed in terrestrial studies. A representative group of individual senescent leaves are air-dried in the laboratory and tied by their petioles to a single line. One end of the line is tied to an identifying tag, and the other end to a flagged washer. Several groups of strings are tied to each washer in this manner: the washer provides the hub, and individual lines the spokes. At each collection interval, one or more lines are snipped from the hub, and measured for decomposition. Subsamples of each line of senescent leaves are then oven dried to determine air-dry/oven-dry ratios. While leaves on a given line usually are weighed individually, mass loss and elemental concentrations are determined for the group to account for any loss of whole leaves. The same care needs to be taken as outlined in the litterbag approach, where leaf litter composition is reflective of the stand under study.

Tethered leaf studies are most useful in studying the early stages of decomposition, thus length of study is not as important as with litterbag approaches. As leaves begin to fragment (a common occurrence early in the decomposition process) this technique will over-estimate decomposition rates relative to the litterbag approach: bag mesh will retain large leaf fragments that would otherwise be lost with the tethered leaf method. Because of the large influence of comminution on estimated decomposition rates for thin or easily fragmented leaves, this method may best yield insights into leaf litter quality for thicker leaves. Studies have shown that small, litter-feeding invertebrates have ready access to litter in litterbags with mesh sizes as low as 1.5 mm (Scowcroft et al. 2000). However, the tethered leaf approach allows for leaf consumption by macroinvertebrates such as crabs and snails, whose access would otherwise be restricted by mesh bags (McKee and Faulkner 2000). Litter is also in direct contact with forest floor, removing methodological artifacts such as changes in forest floor temperature and moisture status.

8.2.4 *Cohort Layered Screen*

A fourth approach to estimating aboveground leaf litter decomposition is the cohort layered window screen method, or litter sandwich method. With this method, layers of mesh screen are used to separate successive layers of litter on the forest floor; leaf litter then decomposes *in situ*.

The cohort layered screen method is applied to long-term decomposition studies, typically three or more years in duration, and is described in detail elsewhere (Binkley 2002). Following major annual litterfall, a layer of window screen is placed over the forest floor. Typically, 1 x 1 m fiberglass or aluminum window screening with a mesh size of 2–3 mm is used. The screen size will depend on the size of the stand sampled, and mesh size will vary with the specific ecosystem under study (see discussion of mesh size under litterbags section). Fiberglass screen is recommended over aluminum if any chemical or constituent properties will be analyzed as well. Following each subsequent annual litterfall for the duration of study, another layer of screen is placed directly over the screen from the previous year. After a given sampling period, subsamples of the original screen can be cut from the original to obtain data while allowing the experiment to continue. Subsamples are collected, weighed, and oven-dried. These are compared with stand level estimates of litterfall for the year in question.

While the litterbag and tethered leaf methods raise concerns about representative leaf quality, the cohort method applies a realistic input of litter species and components, providing the entire litter input for decomposition. It is relatively easy to monitor, as monetary and material resources for preparation and collection are both low, and has been found to represent litter dynamics in the forest floor better than litterbag studies. Such litter sandwiches integrate a large portion of the forest floor, especially for long-term studies. However, the cohort layered screen method also excludes certain macrofauna that are blocked from access to leaf litter by the mesh screen, and can alter the forest floor microclimate.

8.3 Summary

All methods for quantifying litter decomposition suffer from the same inability to separate decomposition losses from leaching and comminution. The distinction is important, but rarely addressed, because the former results in C return to the atmosphere as CO_2 while the latter two processes bring detritus into the soil food web where C may or may not return to the atmosphere. Decomposition constants derived from the litterbag and cohort approaches are fundamentally difficult to scale to the stand because the micro-environment created by both methods and exclusion of organisms will create artifacts. The litterbag approach has the additional problem of accurately representing the forest floor matrix in each litterbag. Comminution losses out of litterbags or through window screen of the cohort approach are of smaller concern than with the tethered litter approach where a break at the petiole is interpreted as decomposition. For all methods, sufficient replication is required to accurately estimate stand level decomposition, and site selection for deployment should represent the various microsites likely to be encountered in the forest stand being examined. This becomes exceedingly difficult to do in diverse stands with complex microtopography. There is also the question of how to capture the decomposition of older material, especially the more decomposed Oa/Oe horizon materials.

For these reasons, it is difficult to impose a "one size fits all" strategy for estimating litter decomposition. One must fully consider the strengths and limitations of each method as it applies to the goals of the study. Since litterfall and forest floor mass will be collected at landscape-scale monitoring sites, one might assume that the mass balance approach would be an easily applicable model, incurring little additional expense. However, as noted earlier, this approach can be imprecise at a scale of annual resolution. If higher resolution is required (e.g., decomposition of branches versus leaves), or forest floor mass is dynamic (e.g., young stands following fire), the cohort layered screen could be used at reasonably small expense. Exclusion of macrofauna, though, could lead to an underestimation of decomposition rates. A realistic approach may be to pair a combination of these techniques as required by various characteristics of each site.

Literature Cited

Binkley D (2002) Ten-year decomposition in a loblolly pine forest. Canadian Journal of Forest Research 32 (12): 2231–2235

Bohlen PJ, Parmalee RW, McCartney DA, Edwards CA (1997) Earthworm effects on carbon and nitrogen dynamics of surface litter in corn agroecosystems. Ecological Applications 7 (4): 1341–1349

Bubb KA, Xu ZH, Simpson JA, Saffigna PG (1998) Some nutrient dynamics associated with litterfall and litter decomposition in hoop pine plantations of southeast Queensland, Australia. Forest Ecology and Management 110: 343–352

Coleman DC, Blair JM, Eliott ET, Freckman DW (1999) Soil invertebrates. In: Robertson GP, Bledsoe CS, Coleman DC, Sollins P (Eds.) Standard soil methods for long-term ecological research. Oxford University Press, New York, pp. 349–377

Dechaine J, Ruan H, Sanchez de Leon Y, Zou X (2005) Correlation between earthworms and plant litter decomposition in a tropical wet forest of Puerto Rico. Pedobiologia 49 (6): 601–607

Giardina CP, Ryan MG, Hubbard RM, Binkley D (2001) Tree species and soil textural controls on carbon and nitrogen mineralization rates. Soil Science Society of America Journal 65: 1272–1279

Gaudinski J, Trumbore S, Davidson E, Zheng S (2000) Soil carbon cycling in a temperate forest: radiocarbon-based estimates of residence times, sequestration rates and partitioning of fluxes. Biogeochemistry 51: 33–69

Hanson PJ, Swanston CW, Garten CT (2005) Reconciling change in Oi-horizon carbon-14 with mass loss for an oak forest. Soil Science Society of America Journal 69: 1492–1502

Harmon ME, Lajtha K (1999) Analysis of detritus and organic horizons for mineral and organic constituents. In: Robertson GP, Bledsoe CS, Coleman DC, Sollins P (Eds.) Standard soil methods for long-term ecological research. Oxford University Press, New York, pp. 143–165

Harmon ME, Nadelhoffer KJ, Blair JM (1999) Measuring decomposition, nutrient turnover, and stores in plant litter. In: Robertson GP, Bledsoe CS, Coleman, DC Sollins P (eds.) Standard soil methods for long-term ecological research. Oxford University Press, New York, pp. 202–240

Hedin LO (2000) Deposition of nutrients and pollutants to ecosystems. In: Sala OE, Jackson RB, Mooney HA Howarth RW (eds.) Methods of ecosystem science. Springer-Verlag, New York, pp. 265–276

Kelly JM, Beauchamp JJ (1987) Mass loss and nutrient changes in decomposing upland oak and mesic-hardwood leaf litter. Soil Science Society of America Journal 51: 1616–1622

Kirschbaum MUF (1995) The temperature dependence of soil organic matter decomposition, and the effect of global warming on soil organic C storage. Soil Biology and Biochemistry 27: 753–760

McKee KL, Faulkner PL (2000) Restoration of biogeochemical function in mangrove forests. Restoration Ecology 8 (3): 247–259

Meentemeyer V (1978) Macroclimate and lignin control of litter decomposition rates. Ecology 59: 465–472

Olsen JS (1963) Energy storage and the balance of producers and decomposers in ecological systems. Ecology 44: 322–331

Peterjohn WT, Melillo JM, Steudler PA, Newkirk KM, Bowles FP, Aber JD (1994) Responses of trace gas fluxes and N availability to experimentally elevated soil temperatures. Ecological Applications 4: 617–625

Raich JW, Nadelhoffer KJ (1989) Belowground carbon allocation in forest ecosystems: Global trends. Ecology 70: 1346–1354

Robertson GP, Paul EA (1999) Decomposition and soil organic matter dynamics In: Sala OE, Jackson RB, Mooney HA, Howarth RW (eds.) Methods of ecosystem science. Springer-Verlag, New York, pp. 104–116

Schlesinger WH (1997) Biogeochemistry, an analysis of global change. 2nd edn. Academic Press, New York.

Scowcroft PG, Turner DR, Vitousek PM (2000) Decomposition of *Metrosideros polymorpha* leaf litter along elevational gradients in Hawaii. Global Change Biology 6: 73–85

Vitousek PM, Turner DR, Parton WJ Stanford RL (1994) Litter decomposition on the Mauna Loa environmental matrix, Hawaii: patterns, mechanisms, and models. Ecology 75(2): 418–429

Xu XN, Hirata E (2002) Forest floor mass and litterfall in *Pinus luchuensis* plantations with and without broad-leaved trees. Forest Ecology and Management 157: 165–173

Chapter 9
Measuring the Decomposition of Down Dead-Wood

Neal A. Scott and Sandra Brown

Abstract Down and dead-wood plays an important ecological role in forest ecosystems, and the decomposition of this material may contribute significantly to forest net ecosystem production. Dead-wood decomposition can be measured in three ways: measuring variability in density over a chronosequence of dead-wood, measuring density changes during a time series of decomposition, and measuring dead-wood respiration rates in conjunction with estimates of dead-wood mass. For sites selected for intensive carbon cycle studies, we suggest a two-prong approach to measuring down dead-wood decomposition. First, initial estimates should be based on a chronosequence of samples stratified by size (1–10 cm diameter, >10 cm diameter). Radiocarbon measurements of the wood in different growth increments should be used to age the material if accurate stand history (mortality) data are not available. Second, we suggest establishing a time-series decomposition experiment to quantify rates of carbon release during the decomposition process. These samples should be stratified by diameter (as above) and by length (approximately 10-times the diameter). This combination of approaches will yield both short- and long-term values for the contribution of dead wood to net carbon sequestration in forest ecosystems, including estimates of associated uncertainty.

Keywords Carbon storage, chronosequence, down dead-wood, decomposition, net ecosystem productivity

---

N.A. Scott
Woods Hole Research Center, Falmouth, MA, Current address: Department of Geography, Queen's University, D201 Mackintosh-Corry Hall, Kingston, ON K7L 3N6, Canada
E-mail: scottn@queensu.ca

S. Brown
Winrock International, Ecosystem Services Unit, 1621 N. Kent Street, Suite 1200, Arlington, Virginia 22209
E-mail: sbrown@winrock.org

C.M. Hoover (ed.) *Field Measurements for Forest Carbon Monitoring*,

9.1 Introduction

Dead-wood, both fallen and standing, plays an important ecological role in forests. In addition to being a potentially large carbon pool with relatively long turnover times (e.g. Harmon and Hua 1991, Currie et al. 2003, Ranius et al. 2003), dead-wood can help reduce rates of erosion, store large amounts of nutrients and water, provide energy and nutrients to soil micro-and macrorganisms, and serve as sites for seedling development (e.g. Lutz 1940, Anderson et al. 1978, Swanson and Lienkaemper 1978, Franklin et al. 1981, Frankland et al. 1982, Davis 1983, Harmon et al. 1986). Many of the important ecological roles of dead-wood are summarized in Harmon et al. (1986). At landscape-scale monitoring sites, we need to quantify the amount of carbon stored in dead-wood and estimate dead-wood decomposition rates to better understand the short- and long-term contribution of dead-wood to the carbon balance of forests. This is particularly important in managed forests where detrital inputs can be highly variable and management-dependent (Grove 2001, Currie et al. 2003, Ranius et al. 2003).

Several factors interact to control the decomposition of dead-wood. Key factors include moisture, wood quality (density, chemistry), size (both diameter and length), and position (contact with the ground). These factors are reviewed in more detail elsewhere (Harmon and Sexton 1996, Harmon et al. 1999), and will not be discussed here except when considering how to establish dead-wood decomposition experiments at landscape-scale monitoring sites.

Most examples of dead-wood decomposition experiments have utilized one of three complementary approaches. Perhaps the most common approach is to sample a chronosequence of dead-wood samples where the age of mortality is known (e.g. Foster and Lang 1982, Yatskov et al. 2003). By measuring variability in wood density at different stages of decomposition, one can infer rates of decomposition by substituting space for time. Age of mortality can be inferred from past disturbance information (e.g. insect outbreaks, fire), seedling age, fall scars, or stumps. Alternatively, the age of death can be estimated by measuring the radiocarbon content of the outer rings (assuming they are still present). This approach can give relatively precise ages on dead-wood where mortality occurred after 1960, the last year of atmospheric bomb testing that introduced a ^{14}C signal into the atmosphere (see Trumbore et al. 1995). An alternative to the chronosequence approach is the time-series approach, where fresh dead-wood samples are placed in the field and their decomposition is monitored over time (e.g. Stone et al. 1998, Fraver et al. 2002). This approach yields more information on the temporal dynamics of wood decomposition, but takes several years before decomposition rates can be calculated. Both of these methods involve measurements of changes in wood density over time, with the chronosequence approach substituting space for time. Dead-wood decomposition can also be estimated by measuring rates of dead-wood respiration across a range of decay classes, estimating annual dead-wood respiration using an empirical relationship between respiration and temperature, then estimating annual decomposition rates based on dead-wood pool sizes (e.g. Chambers et al. 2001, Bond-Lamberty et al. 2003). This method measures decomposition related to heterotrophic activity, but does not account for key dead-wood decomposition processes of dissolved organic carbon leaching and fragmentation (Harmon and Hua 1991, Currie et al. 2003). The relative merits of these different approaches are discussed in more detail in Harmon and Sexton (1996).

For landscape-scale monitoring sites, we suggest a two-pronged approach to estimating wood decomposition. First, we recommend establishing a time-series experiment with fresh dead-wood (fine and coarse) that will be used to quantify wood decomposition over time. Then, to obtain initial estimates of wood decomposition, we recommend using a chronosequence approach where ^{14}C in the outer rings is used to estimate age of death if detailed mortality information is not available to age the dead-wood. Radiocarbon measurements should be accompanied by measures of wood density on the different decay classes.

9.2 Sampling Design/Issues

9.2.1 Approaches

For wood decomposition experiments based on a time-series, a method similar to litterbags can be used to measure decomposition rates. However, due to the slower decomposition rates and integrity of the material, the litter (wood) does not have to be confined in litterbags. For fine woody detritus (>1 and <10 cm diameter), whole pieces of wood will be collected at different points in time and weighed to determine mass loss at each sampling period. A similar method will be used for coarse wood (>10 cm), except that the pieces will be sub-sampled in the field due to their large size, and the sub-samples will be dried and weighed to determine changes in wood density over time.

It is not expected that valid chronosequences (e.g. Chen et al. 2001) of dead-wood will be available at the landscape-scale monitoring sites. However, it is desirable to obtain initial estimates of wood decomposition rates (including coarse roots) that do not require waiting ten years. If previous disturbances have occurred near the monitoring sites, we suggest obtaining some initial estimates of dead-wood decomposition by sampling the density of material that has decomposed over different time scales, then using ^{14}C techniques (e.g. Trumbore et al. 1995) to quantify the age of death. While ^{14}C analyses are not inexpensive (on the order of $100 to $150 per sample), they provide robust estimates of sample age, particularly over the last 40 + years due to the ^{14}C enrichment of the atmosphere from bomb testing. If disturbance history is well-documented for the site, mortality dates can be obtained without the use of the more expensive radiocarbon methods.

9.2.2 Size

Size, both diameter and length (since wood decomposition is often initiated at the ends of the debris), is a key factor influencing rates of decomposition. For the purposes of landscape-scale monitoring we suggest using diameter classes of >1 and <10 cm to delineate fine woody detritus, and diameters >10 cm for coarse woody detritus. As for the length of pieces used for time-series wood decomposition experiments, we suggest using pieces where the length is about ten times the diameter of the piece (Harmon and Sexton 1996). The pieces need to be long enough to allow for removal of multiple cross sections when sampled. These lengths do not have to be exact – a tolerance of ± 20% is probably acceptable over all the samples. We also suggest distributing the sizes (diameters) as much as possible throughout the range needed for a given site. Harmon et al. (1999) suggest using a geometric distribution of diameters (e.g. 1, 2, 4, 8, 15, and 30 cm), with a similar number of samples for each size. As with the length, some tolerance in selection of diameters is acceptable. Regardless of what pieces are selected, an important thing to remember is to ***record initial conditions/information for each sample.***

9.2.3 Sampling Intervals

Decomposition of coarse wood can vary from decades to centuries depending on the type of wood and the environment. It makes little sense, therefore, to prescribe a sampling interval for coarse wood. Harmon et al. (1999) suggest sampling intervals of about 10% of expected life span. For fine wood material, expected life span is probably 5–10 years. Based on these estimates, we suggest fixed sampling intervals of 6 months (for the first two years), and then annually thereafter. For coarse wood material, we recommend measuring decomposition at 15 points during the decay process. Individual sites will have to decide on the best sampling intervals using *a priori* knowledge about wood decomposition rates. For example, if we assume a lifespan of 25 years we suggest annual sampling for the first five years, then biennial sampling thereafter.

9.2.4 Position

Position of the dead-wood has a large influence on rates of wood decomposition (Harmon and Sexton 1996). While it might be prudent to establish decomposition experiments with material that is not contacting the forest floor, we suggest that all samples be placed on the ground and in contact with the forest floor for consistency. Samples should be adequately marked with string/tags to ensure they can be relocated even when covered by annual litterfall.

9.2.5 Species

Wood quality also has a large impact on wood decomposition rates. While it would be useful to include all dominant species at all sites, this may not be practical in

Table 9.1 Decay resistance of North American genera

Very resistant	Resistant	Moderate resistance	Low resistance
Maclura	Calocedrus	Gleditsia	Abies
Morus	Castanea	Larix	Acer
Robinea	Catalpa	Lithocarpus	Alnus
Taxus	Chamaecyparis	Pinus (old yellow)	Betula
	Juglans	Pinus (white)	Carya
	Juniperus	Pseudotsuga	Celtis
	Quercus (white)	Quercus	Fagus
	Prunus		Fraxinus
	Sassafras		Liquidambar
	Sequoia		Liriodendrum
	Sequoiadendron		Magnolia
	Taxodium		Nyssa
	Thuja		Picea
			Plantanus
			Populus
			Pinus (young)
			Quercus (black)
			Salix
			Tilia
			Tsuga
			Ulnus

Based on Table 8 in Harmon and Sexton (1996)

terms of both cost and time. At sites dominated by one species, that species should be the only one used for measuring wood decomposition rates. When there are two dominant species present, one species should be selected for wood decomposition measurements unless *a priori* information suggests large differences in wood decomposition rates. Where multiple species are present with (potentially) large differences in wood decomposition rates, include species with the (expected) slowest and fastest decomposition rates in any wood decomposition experiments. This division can often be done at the level of genus (Table 9.1).

9.3 Fine Wood Decomposition Time Series

9.3.1 Initial Conditions

One of the most important steps in establishing a field-based wood decomposition experiment is to record ample information about the samples before the experiment begins. This includes total diameter, length, radial increments (heartwood, sapwood and bark), total volume, bark cover, and species. These data will be crucial to help interpret results from the experiment.

9.3.2 Sample Preparation

Sources of material for fine wood decomposition experiments can include recently downed trees, or fresh material can be obtained by harvesting a tree at each site. When preparing individual pieces, it is best to start with long pieces and then cut sub-samples. Use the diameter classes described previously (1, 2, 4, and 8 cm), with equal numbers of samples in each diameter class. If starting with already dead material, remove the first ~5 cm from the ends before starting to remove samples. Then, starting at one end, cut a sample of appropriate length (~10 times the diameter). At the same time, remove a small sub-sample to be used to measure initial moisture content. Given the high variability of moisture content, it is important to have a moisture sample for each sample to be placed in the field. Weigh each piece, then label with a metal tag. Place the sub-sample for moisture determination in a plastic bag labeled with the number on the metal tag. Once back in the lab, this sub-sample is dried to a constant mass at 55°C. This same sample can also be used for initial chemistry measurements if required. Place the tagged samples around the site, making sure that they are all in contact with the forest floor. Prepare a site map showing where the samples are located to facilitate retrieval at a later time. **Note all initial conditions (length, species, diameter at center of piece, other characteristics [e.g. knots etc.])**.

9.3.3 Sample Replication

Ideally, if these data were to be used to compare decomposition rates across landscape-scale monitoring sites, it would be good to establish multiple 'sites' at each site to avoid pseudoreplication (Hurlbert 1984). However, this would add significantly to both the cost and time requirements. Therefore, use one 'site' at each landscape-scale monitoring installation, with several (at least 4) replicate samples collected at each sampling interval. For example, for a given diameter (1 cm), there will be seven or more sample points in the time series, and four pieces for each diameter class should be collected at each sample interval. This means that initially, a minimum of 28 pieces of wood with diameter of ~1 cm will need to be cut assuming a 5-year experiment. Ideally, additional pieces (about 36 wood samples in each diameter class) should be prepared in case additional sampling is required beyond five years. If the scope of the experiment is to be expanded, we would suggest adding position (e.g. some samples not in contact with the forest floor along the entire length) as an experimental factor. Then, if the scope is to be expanded even further, multiple independent 'sites' at each landscape-scale monitoring site could be established to facilitate more robust between-site statistical comparisons of dead-wood decomposition rates.

9.3.4 Sample Collection

At each sample time (6, 12, 18, 24, 36, 48, 54 months), collect four samples of each diameter class. If the entire sample is collected, either weigh the sample in the field

OR place it in a plastic bag and return to the lab for weighing. Once the wet mass has been measured, either dry the entire sample (55°C to constant mass) OR collect 4 evenly-spaced sub-samples from each piece, weigh the wet sub-samples, then dry to constant mass. At the time of collection, make a visual assessment of structural (volume) changes due to fragmentation. **It should NOT be necessary to measure volume changes with the fine material as long as the entire sample can be weighed wet.** Prior to drying, be sure to remove any attached pieces of mineral or organic matter. Dry all (sub) samples at 55°C to a constant mass and weigh. If there is significant evidence of mineral-soil contamination of the (sub)samples, then the sub-sample should be ashed at 500°C in a muffle furnace **after** it has been dried at 55°C and weighed.

9.3.5 Calculations

Dry weights of samples are determined by multiplying the wet weight by the dry weight: wet weight ratio of the sub-sample (for samples where sub-samples are removed). If the entire piece is dried and weighed, no correction is required. For samples that decompose on the surface of the soil, contamination with mineral soil is very small, so correction for ash content is probably not required. Depending on the kinetics, decomposition rates can be estimating using regression techniques and either a zero-order equation (Eq. 9.1) or a first order decay equation (Eq. 9.2), where A_t is sample mass at time t, A_o is initial sample mass, t is time, and k is the decomposition rate. Data can be linearized using log transformations if necessary.

$$A_t = A_o - kt \tag{9.1}$$

$$A_t = A_o e^{-kt} \tag{9.2}$$

9.4 Coarse Wood Decomposition Time Series

While fine wood decomposition experiments are relatively easy to establish, coarse wood decomposition experiments, if done correctly, require a significant amount of work and heavy equipment to move logs. Whether this is feasible at landscape-scale monitoring sites will depend on both cost and accessibility – these will likely be resolved on a site-by-site basis.

9.4.1 Initial Conditions

As with fine wood, one of the critical steps in a time series wood decomposition experiment is to record ample information about the initial conditions. This includes total diameter, length, radial increment (heartwood, sapwood, bark), total volume, bark cover, and species. Bark cover can be estimated by sliding a frame (rectangular) over the outside of the log and estimating the proportion of the log sample covered by bark in each rectangle. These data will be crucial for estimating volume changes and interpreting results. It is also useful to record the depth of the pith, as this serves as a useful reference point as the piece fragments (Harmon et al. 1999). Selection of species follows similar criteria to those used for fine wood decomposition.

9.4.2 Sample Preparation

Sources of material for coarse wood decomposition experiments can include recently downed trees, or fresh material can be obtained by harvesting a tree. When preparing individual sample pieces, it is best to start with long pieces and then cut sub-samples. Use the diameter classes described previously (15 and 30 cm), with equal numbers of samples in each diameter class. If starting with already dead material, remove the first ~5 cm at each end before starting to remove samples. Starting at one end, cut a sample of appropriate length (about 10 times the diameter), then remove a cross-sectional sub-sample to be used to measure initial moisture content and wood density. Ideally, for the initial samples, a sub-sample is removed from both ends of each sample log. Given that the sample logs are too large to be weighed in the field, it is **critical** to measure the initial volume of each sample log. At a minimum, this is accomplished using measures of total length and the diameter at both ends and in the middle. Log volume is estimated using Newton's formula (Eq. 9.3).

$$V = L * (A_b + 4A_m + A_t) / 6 \qquad (9.3)$$

where V is log volume, L is log length, A is the cross-sectional area of the log at the bottom (b), middle (m), and top (t). If the logs are non-circular, both minimum and maximum diameter should be measured at each location on the log. Diameters can be measured using either a diameter tape or a pair of calipers (calipers are probably easier unless the log is too large). To calculate initial mass, log volume is multiplied by wood density, which is determined using the cross-sectional sub-samples (see *Estimating Wood Density*). Once the pieces are cut, label them and place the tagged samples around the site, making sure that they are all in contact with the forest floor.

9.4.3 Replication

As with the fine woody material, we do not suggest establishing multiple 'sites' at each landscape-scale monitoring site. As discussed previously, time intervals for sampling dead-wood decomposition will vary from site to site, but here we provide an example assuming 25 years of decay (to obtain 15 sample points). Sampling times would be at 1, 2, 3, 4, 5, 7, 9, 11, 13, 15, 17, 19, 21, 23, and 25 years. For coarse wood, sampling of three replicate logs at each time is probably adequate. This means a total of 90 logs have to be prepared for this experiment, assuming that each log is destructively sampled at each sample time (and two diameter classes are used, as described above).

9.4.4 *Sample Collection*

At each sample time, collect three replicate sample logs from each diameter class. For each log, estimate the log volume using similar methods to those used for the initial sample (measure diameter at both ends and in the middle at a minimum). Quantify (changes in) bark cover, and note any structural changes (i.e. fragmentation) in the log. If the volume has changed markedly from the original sample, corrections for fragmentation may be required. Remove at least four evenly spaced cross-sectional sub-samples with a chain saw from each log for density determinations.

9.4.5 *Estimating Wood Density*

Volume and mass of the cross-sectional sub-samples must be measured in order to estimate changes in wood density and other mass-loss processes (e.g. fragmentation) during decomposition. Estimating volume can be difficult, and various methods have been used depending on the decay status of the log (e.g. Sollins et al. 1987, Harmon 1992). For the most precise estimates of dead-wood decomposition, the sample can be split into various components with different decay rates (e.g. bark, sapwood, heartwood) and the volume estimated appropriately (see Fig. 9.1; see Harmon et al. 1999 for additional details). Alternatively, volume of whole samples (or parts) can be measured by displacement (samples can be wrapped tightly with plastic wrap to minimize water absorption). For heavily decomposed samples, the only possible way to measure volume is to measure the void produced by removing a subsample from the piece of wood. While combining the various parts of the wood into a single sample introduces some errors, it reduces the time required for sampling considerably, and may therefore be appropriate for the landscape-scale monitoring sites. A final decision will rest with the investigators and may be influenced by the relative importance of dead wood to the C budget of a particular forest.

Once the volume has been measured and the sample weighed, subsamples must be collected for moisture determination. Alternatively, the entire layer can be chopped up and dried. Remove any attached pieces of mineral or organic matter prior to drying. Dry all samples at 55°C to a constant mass and weigh. If there is significant evidence of mineral soil contamination of the (sub) samples, then samples should be ashed at 500°C in a muffle furnace **after** drying at 55°C.

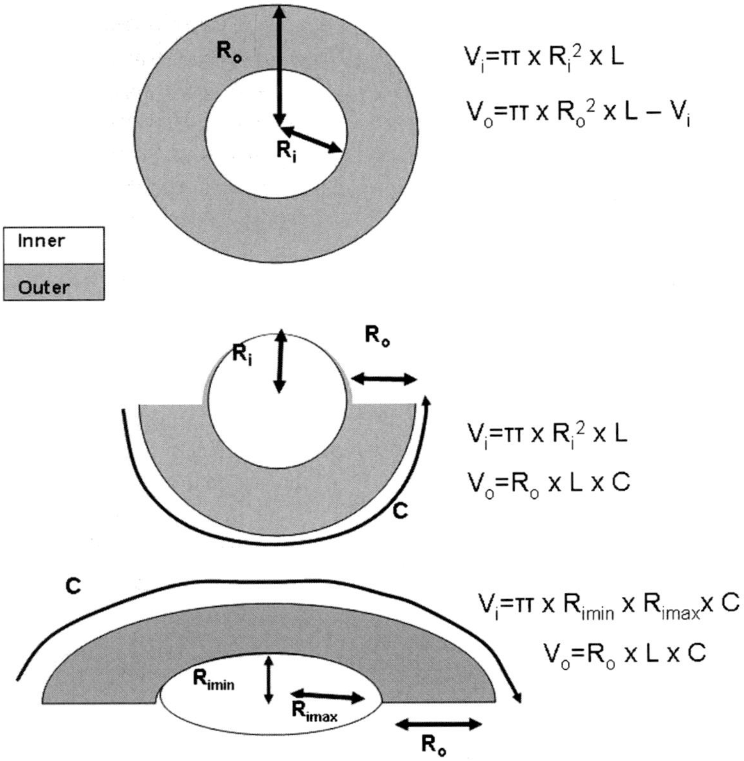

Fig. 9.2 Measurements and formulas to determine volume of different layers contained in the cross-sectional sub-samples of coarse dead-wood. R is the radial dimension, C is the length along the circumference. L is the longitudinal dimension and is not shown. Subscripts indicate whether dimension is inner (i) or outer (o) layer or the minimum **(min)** or maximum **(max)** axis (From Harmon et al. 1999)

9.4.6 Calculations

If a layer sub-sample is dried, calculate total layer oven-dry mass by multiplying total layer fresh mass by the ratio of sub-sample oven-dry weight:fresh weight. Wood density can then be calculated by dividing oven-dry mass by total volume, giving density in units of g/cm^3. If the total layer is dried, simply divide the final oven-dry weight by the total volume. If significant mineral-soil contamination was present and the sample had to be ashed, correct the final oven-dry mass using the estimated ash content calculated by multiplying the pre-ash weight by the ratio of post-ash weight:pre-ash weight.

To estimate decomposition rates, non-linear regression techniques can again be used with various decay equations (Eqs. 9.1 and 9.2). If it is not possible to separate heartwood from sapwood, then a two-pool exponential decay model (Eq. 9.4) may be most appropriate for describing decomposition kinetics. In this model, A_1 and A_2

are pool sizes of the two different components (e.g. heartwood and sapwood), and *k1* and *k2* are the decay rates for these different components.

$$A_t = A_1 e^{-k1t} + A_2 e^{-k2t} \tag{9.4}$$

If desired, investigators may want to collect digital photos of the samples at each sampling interval to keep a visual record of the sample decay sequence.

9.5 Chronosequence Approach

With the chronosequence method, measurements of log volume and wood density are identical to the methods described for measuring decomposition of coarse dead-wood material over a time series. Ideally, sample logs can be selected in such a way to represent all of the major decay classes (e.g. Arthur et al. 1993, Grove 2001, Fraver et al. 2002). As with the time series approach, we suggest trying to find at least three replicate logs in each decay class. Initial log volume and density must be estimated using site-specific wood density information and reconstructions of log volume – this is one source of uncertainty in the chronosequence approach. If the date of death of a tree is not known based on site history, it is possible to estimate the date of death (if after 1960) by collecting a small sample of the outer 2–3 ring increments and having them analyzed for radiocarbon. Details on collection procedures, sample quantities, and preparation are not included here, but should be obtained in consultation with experts accustomed to working with radiocarbon measurements. Examples of the chronosequence approach to quantifying dead-wood decomposition can be found in Chen et al. (2001) and Yatskov et al. (2003).

9.6 Measuring Down Dead-Wood Decomposition in Harvested Sites

Most of the methods described above apply to measuring wood decomposition rates of single logs or branches that fall to the forest floor. In forests that have undergone recent harvesting activity, particularly with harvests that distribute debris throughout the forest, it is common to find dead-wood in large piles as opposed to single pieces lying on the forest floor (e.g. Scott et al. 2004). Within these piles, you have wood exposed to a wide range of environmental conditions that are difficult to emulate in a sampling scheme such as that described above. Different methods may therefore be required to measure down dead-wood decomposition in these managed forests.

One possible approach is to create artificial slash piles that can be periodically reweighed to quantify mass loss over time (see Fig. 9.2). The size and amount of material used in the pile will depend on measurements of dead-wood produced by

Fig. 9.2 Reconstructed slash piles at Howland Forest, Maine, used to quantify decay rates of down dead-wood produced during a shelterwood harvest (Photograph by N. Scott)

the harvest. In this example (Fig. 9.2), we created experimental plots that were roughly the width of the logging roads created during the harvesting operation (3.4 × 2.1 m (area of 7.1 m^2)). Plot area is delineated with 2 mm mesh nylon screening to facilitate resampling. Each slash pile contained roughly 50 kg of slash initially, the amount based on prior sampling of woody debris standing stocks after the harvest. During the first two years, mass remaining (determined by weighing the entire slash pile) was determined every six months. After that time, sampling occurred every year. At each sample period, pieces of wood are placed in a large trash can and weighed in the field with a suspended scale. Samples are stratified by size (<1 cm diameter, 1–5 cm diameter, >5 cm diameter). We included additional smaller size-classes to reflect the large amount of relatively fine material in the samples. Subsamples approximating 3% of each can of material are collected for moisture content. These samples are dried at 55°C and weighed. After the piles are weighed, they are returned to the pile in roughly the same location as they were prior to sampling. While some material is lost at each measurement period, we avoid errors associated with volume determinations.

Literature Cited

Arthur MA, Tritton LM, Fahey TJ (1993) Dead bole mass and nutrients remaining 23 years after clear-felling of a northern hardwood forest. Canadian Journal of Forest Research 23: 1298–1305

Anderson NH, Sedell JR, Roberts LM, Triska FJ (1978) The role of aquatic invertebrates in processing wood debris in coniferous forests streams. American Midland Naturalist 100: 64–82

Bond-Lamberty D, Wang C, Gower ST (2003) Annual carbon flux from woody debris for a boreal black spruce fire chronosequence. Journal of Geophysical Research 107: 8220, doi:10.1029/2001JD000839, 2003. [printed 108(D3), 2003]

Chambers JQ, Schimel JP, Nobre AD (2001) Respiration from coarse wood litter in central Amazon forests. Biogeochemistry 52: 115–131

Chen H, Harmon ME, Griffiths RP (2001) Decomposition and nitrogen release from decomposing woody roots in coniferous forests of the Pacific Northwest: a chronosequence approach. Canadian Journal of Forest Research 31: 246–260

Currie WS, Yanai RD, Piatek KB, Prescott CE, Goodale CL (2003) Processes affecting carbon storage in the forest floor and in downed woody debris. In: Kimble JM, Heath LS, Birdsey RA, Lal R (Eds.) The potential of U.S. forest soils to sequester carbon and mitigate the greenhouse effect. CRC Press, Boca Raton, FL

Davis JW (1983) Snags are for wildlife. General Technical Report RM-99. U.S. Forest Service, Rocky Mountain Research Station, Flagstaff, Arizona

Foster JR, Lang GE (1982) Decomposition of red spruce balsam fir boles in the White Mountains of New Hampshire. Canadian Journal of Forest Research 12: 617–626

Frankland JC, Hedger JN, Swift MJ (1982) Decomposer Basidiomycetes-their biology and ecology. Cambridge University Press, Cambridge, MA, 355p

Franklin JR, Cromack K Jr, Denison W, McKee A, Maser C, Sedell J, Swanson F, Juday G (1981) Ecological characteristics of old-growth Douglas fir forests. Gen Tech Rep. PNW-118. USDA Forest Service Pacific Northwest Forest Experiment Station, Portland, OR, 48 p

Fraver S, Wagner RG, Day M (2002) C dynamics of coarse woody debris following gap harvesting in the Acadian forest of central Maine, U.S.A. Canadian Journal of Forest Research 32: 2094–2105

Grove SJ (2001) Extent and composition of dead wood in Australian lowland tropical rainforest with different management histories. Forest Ecology and Management 154: 35–53

Harmon ME (1992) Long-term experiments on log decomposition at the H.J. Andrews Experimental Forest. USDA Forest Service, General Technical Report PNW-280, Portland, OR

Harmon ME, Franklin JF, Swanson F, et al. (1986) Ecology of coarse woody debris in temperate ecosystems. Advances in Ecological Research 15: 133–302

Harmon ME, Hua C (1991) Coarse woody debris dynamics in two old-growth ecosystems. Bioscience 41: 604–610

Harmon ME, Nadelhoffer KJ, Blair JM (1999) Measuring decomposition, nutrient turnover, and stores in plant litter. In: Robertson GP, Coleman DC, Bledsoe CS, Sollins P (Eds.). Standard Soil Methods for Long-Term Ecological Research, Oxford University Press, Oxford

Harmon ME, Sexton J (1996) Guidelines for Measurements of Woody Detritus in Forest Ecosystems. Publication No. 20. U.S. LTER Network Office: University of Washington, Seattle, WA, 73 p

Hurlbert SH (1984) Pseudoreplication and the design of ecological field experiments. Ecological Monographs 54: 187–211

Lutz HJ (1940) Disturbance of forest soil resulting from the uprooting of trees. Yale School of Forestry Bulletin 45. Yale School of Forestry, New Haven, CT, 37p

Ranius T, Kindvall O, Kruys N, Jonsson BG (2003) Modelling dead wood in Norway spruce stands subject to different management regimes. Forest Ecology and Management 182: 13–29

Scott NA, Rodrigues C, Hughes H, Lee JT, Davidson EA, Dail DB, Malerba P, Hollinger DY (2004) Changes in carbon storage and net carbon exchange one year after an initial shelterwood harvest at Howland Forest, ME. Environmental Management 33(Suppl. 1): S9–S22

Sollins P, Cline SP, Verhoeven T, Sachs D, Spycher G (1987) Patterns of log decay in old-growth Douglas-fir forests. Canadian Journal of Forest Research 17: 1585–1595

Stone JN, MacKinnon A, Parminter JV, Lertzman KP (1998) Coarse woody debris decomposition documented over 65 years on southern Vancouver Island. Canadian Journal of Forest Research 28: 78–793

Swanson EJ, Lienkaemper GW (1978) Physical consequences of large organic debris in Pacific Northwest streams. Gen. Tech. Rep. PNW-69. USDA Forest Service Pacific Northwest Forest and Range Experiment Station, Portland, OR, 12 p

Trumbore SE, Davidson EA, de Camargo PB, Nepstad DC, Martinelli LA (1995) Belowground cycling of carbon in forests and pastures of Eastern Amazonia. Global Biogeochemical Cycles 9: 515–528

Yatskov M, Harmon ME, Krankina ON (2003) A chronosequence of wood decomposition in boreal forests of Russia. Canadian Journal of Forest Research 33: 1211–1226

Part IV
Measuring Belowground Carbon Pools and Fluxes

Chapter 10
Measuring Forest Floor, Mineral Soil, and Root Carbon Stocks

Andrew J. Burton and Kurt S. Pregitzer

Abstract In forest and wildland ecosystems, forest floor and soil carbon (C) comprise a large C pool that is often of similar magnitude to or greater than aboveground C storage. These C pools often change slowly over time, but they are susceptible to rapid release to the atmosphere following natural or human-caused disturbances. Accurate estimates of these pools are needed both to quantify current ecosystem C storage and to understand the potential for future soil C sequestration or release due to disturbance, reforestation, or global change factors. This chapter describes a field sampling approach for quantifying C stocks in forest floor and mineral soil horizons and in live and dead coarse root biomass. The volumetric pit method described for soil horizons is time consuming and labor intensive, but is known to produce the least bias, especially for soils with high stone and rock content. The method described is designed for upland soils, but alternatives that can be used in wetland soils are listed. Key considerations for any sampling of soil C stocks are obtaining accurate volumes and densities for soil materials sampled and sampling to sufficient depth to obtain the vast majority of organic soil C. Careful sampling of soil C stocks at periodic intervals will provide information that helps us model changes in soil C due to management, disturbance and global change and helps validate rates of net ecosystem C exchange measured by flux towers and other techniques.

Keywords Bulk density, forest floor, soil carbon, root biomass, volumetric pit.

A.J. Burton
School of Forest Resources and Environmental Science, Michigan Technological University, 1400 Townsend Drive, Houghton, MI 49931
E-mail: ajburton@mtu.edu

K.S. Pregitzer
Department of Natural Resources and Environmental Science, University of Nevada, 1000 Valley, Road, Reno, NV 89512
E-mail: ksp@cabnr.unr.edu

C.M. Hoover (ed.) *Field Measurements for Forest Carbon Monitoring*,

10.1 Introduction

Soil carbon (C) stocks are one of the largest C pools in forest and wildland ecosystems. Coarse root biomass also can be a large pool, with values for mature stands typically around 20–30% of the C content in aboveground woody biomass. Accurate estimates of these pools are necessary to quantify current C storage and to understand the potential for factors such as elevated CO_2, climatic change, chronic N deposition, land use change and catastrophic disturbance to alter ecosystem C storage and net C exchange. The sampling procedures for soil C stocks described below are quite labor intensive, requiring at least two volumetric pits per sample plot. However, given the spatial heterogeneity common in most soils, we feel this sampling intensity is near the minimum level required to achieve acceptable precision. Changes over time in soil C stocks often are slow, and without a high level of precision, small changes in these pools will be impossible to detect.

Sampling of initial locations will allow determination of the power the design provides for detecting differences among locations and over time. The sampling scheme can then be adjusted to provide a specified, acceptable level of precision. Soil C stocks should not be remeasured on intervals of less than 5 years. Since C stocks in the forest floor and surface soil can vary seasonally, all sampling should occur at the same time of year, preferably in late summer. The sampling scheme focuses primarily on forest floor and near surface soil increments, where the majority of soil organic C typically is located. Intensive sampling is used to a depth of 70 cm, to adequately sample B horizon accumulations of C. Below 70 cm, sampling can be much less intensive, as most upland soil horizons at these depths will contain little organic C.

10.2 Forest Floor Carbon Content

Forest floor can be sampled using frames of various dimensions; a 30 x 30 cm frame is recommended. A minimum of four randomly selected locations per plot should be sampled. At each location, separate samples should be taken from the Oi (L), Oe (F), and Oa (H) layers using the following standard definitions:

> Oi horizon: Organic material having undergone little or no decomposition (fibric material). On the forest floor this layer consists of freshly fallen leaves, needles, twigs, stems, bark, and fruits. This layer may be very thin or absent during the growing season.
>
> Oe horizon: Partially decomposed litter with portions of plant structures still recognizable (humic material). This horizon occurs below the Oi in the forest floor.
>
> Oa horizon: A layer consisting of well-decomposed organic matter of unrecognizable origin (sapric material).

These forest floor horizons may not all exist at every location.

10.2.1 Forest Floor Sampling and Carbon Content Calculation

1. Materials needed for sampling forest floor horizons include: a square metal frame (typically 30 x 30 cm), a serrated knife, and paper bags for transporting, drying and storing samples.
2. Place the metal frame on the forest floor
3. Use a serrated knife to cut through the forest floor around the inside of the frame prior to removing the layers; any live plant material inside the frame should be removed prior to sampling. Figure 10.1 illustrates the use of a frame and knife for sampling forest floor.
4. Separately remove material from each forest floor horizon described above and place in a paper bag. Samples should be kept cool (<4°C) until they can be returned to the laboratory and processed. For wet samples, it may be necessary to place the paper bags in plastic bags for storage prior to drying.
5. In the laboratory samples should be dried as soon as possible at 70°C for 48 h.
6. After drying, samples should be weighed, ground and analyzed for C and N (g kg^{-1}) using an elemental analyzer (see section 10.3.4)
7. Carbon and N contents (g m^{-2}) for each forest floor horizon can be calculated by multiplying sample mass (kg) by concentration (g kg^{-1}) and dividing by sample area (m^2).

10.3 Mineral Soil Carbon Content

Samples for determining mineral soil C stocks can be obtained using volumetric pits to a depth of 70 cm. These should be taken from beneath two of the forest floor sampling locations per plot. Separate samples should be taken from depth increments

Fig. 10.1 Illustration of the use of a knife and metal frame to cut a block of forest floor material for sampling (Photo by C. Hoover)

of 0–10, 10–20, 20–30, 30–50 and 50–70 cm. Core sampling is commonly used for determining soil C contents, but it is not recommended for universal use. Core sampling is difficult to apply in stony soils, with many holes abandoned due to the inability of the corer to reach the desired depth (Muller and Hamilton 1992). This causes the sample locations to be biased toward the less stony portions of the soil profile (Harrison et al. 2003, Park et al. 2007). Also, as corers are extended deeply into any soil, the potential exists for soil to be pushed in front of or to the side of the corer, resulting in incomplete sampling of the desired depth increment and errors in estimates of coarse fragment fractions, roots and soil bulk density (Page-Dumroese et al. 1999, Park et al. 2007).

10.3.1 Field Sampling of Mineral Soil Horizons

1. Materials needed for sampling mineral soil horizons include shovels and hand trowels for excavating soil materials and paper and plastic bags for storing samples. Heavy duty shovels with long, fairly straight blades (e.g. sharpshooter style) are preferred for their ability to create sample pits with nearly vertical sides.
2. Under two of the forest floor sampling locations per plot, a pit approximately 20 x 20 cm in area should be excavated to 10 cm, with all soil and stones removed.
3. All material removed should be placed in sturdy paper or plastic bags and returned to the lab for analysis. Samples should be double bagged to prevent loss of sample volume due to bag breakage.
4. If the water method is being used to estimate the volume of soil material removed, the volume of the depth increment should be determined prior to extending the pit (see section 10.3.2 below). If polyurethane foam is being used to determine the volume of soil materials removed, then the bottom of the depth increment should be marked with nails inserted into the soil. This will facilitate determination of separate volumes for each depth increment when the foam is used to create a mold of the entire pit volume.
5. The pit can then be sequentially extended for sampling the remaining depth increments to 70 cm. If needed for the deeper depth increments, a board can be placed over the surface of the sampling area to be extended downward, and the pit above that depth can be widened to facilitate excavation of the next depth increment (Fig. 10.2).
6. Beneath 70 cm, additional samples for soil C and N can be made using soil augers, in depth increments of 70–100, 100–130, and 130–160 cm. In most cases, total and fine bulk densities (section 10.3.3.1) will not be determined for these depths. They will be assumed to be the same as that for the 50–70 cm depth, unless obvious differences exist, such as the presence of a fragipan or other compacted layer. In such cases it may be necessary to extend the volumetric

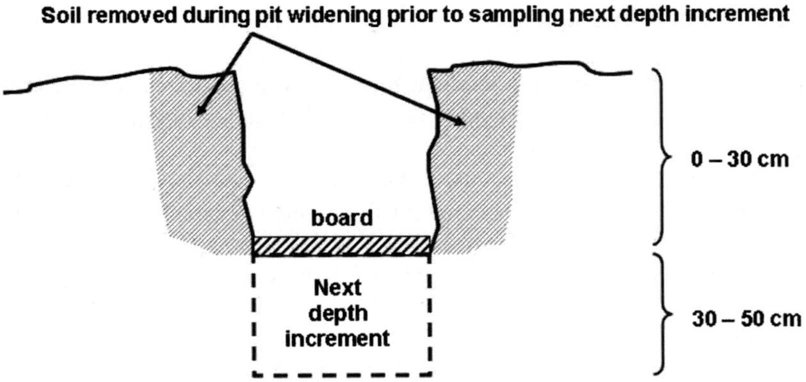

Fig. 10.2 Illustration of pit widening between sampling of the 20–30 cm and 30–50 cm depth increments. A board is used to protect the pit bottom in the sampling area during pit widening. For deeper depths, such width expansion will be necessary to enable working access to the next soil sampling increment (30–50 cm in this illustration)

pit to determine bulk density for these depths. In extremely stony soils it also may be necessary to extend the pit to obtain the necessary samples of the fine soil material from deeper depths.

7. Samples should be kept cool (<4°C) until they can be returned to the laboratory and processed.

10.3.2 Sample Volume Determinations

Quantification of coarse-fragment volume is needed to ensure accurate estimates of soil C stocks per unit area. In stony soils, errors in determination of this volume can greatly affect calculated C content. As a first step toward determining stone and fine soil volumes, the total volume of each depth increment sampled needs to be determined.

10.3.2.1 Volume by Water Method

1. The total volume of each depth increment sampled needs to be determined by placing a plastic bag in the hole and recording the volume of water needed to fill the hole (Howard and Singer 1981) for the depth increment.
2. To avoid puncturing the bag, care should be taken during excavation to remove protruding rocks and roots from the sides of the hole.
3. If the pit is not widened between increments, depth for lower increments can be determined by subtracting the volumes of previous depth increments.

10.3.2.2 Volume by Foam Method

1. At locations where water is not available nearby and it is not feasible to carry water in, expanding polyurethane foam (Muller and Hamilton 1992) can be used for volume determination.
2. When using the foam, a cardboard plate must be placed on the surface with a weight on top to ensure that the foam will continue to expand inside the pit and fill any irregularities.
3. Markers, such as small nails, should be left protruding from the soil to help delineate the depth increments used, as this method will be used only once for the entire pit.
4. Foam should allowed to cure for a minimum of 8 h, and ideally 24 h, after which foam casts can be removed and returned to the lab. Any soil or root materials adhering to the foam cast should be removed.
5. The foam casts are cut into separate in depth increments and sprayed with water resistant clear gloss urethane (Brye et al. 2004) to create a waterproof seal on the cut edges. Volumes for each depth increment are then determined by water displacement.

10.3.3 Separation and Density Determination of Stone and Soil Fractions

In the laboratory, soils should be air-dried if they will not be immediately processed. An oven set to a low temperature (30–35°C) can be used to aid this process. Alternatively, samples can be spread out on trays in a warm, dry area for air drying.

1. For each depth increment stones (>10 mm) are removed by sieving.
2. Subsamples of stones should have both their weight and volume (by displacement) determined in order to estimate particle density, (D_{St}).
3. Organic material not passing the 10 mm screen will be placed with the rest of the sample, unless it is a root > 10 mm in diameter.
4. Roots > 10 mm in diameter are accounted for by the coarse root biomass estimate described in section 10.4.1. However, the diameter and length of each should be measured, to allow subtraction of their volume (V_{CR}) from the total sample volume. Coarse root fragments should be dried (70°C for 48 h), ground and analyzed for C and N to provide estimates of coarse root C content to be applied to coarse root biomass estimates described in section 10.4.1.
5. The remaining portion of each sample should be sieved using a 2 mm screen. Materials not passing through the 2 mm screen are divided into stones and organic material. Note that the organic material will include some roots < 10 mm in diameter (fine roots). Soil aggregates should be forced through the screen by hand or using a rubber-tipped pestle. Aggregates can be destroyed using a rolling pin on a hard flat surface if necessary, then passed through the screen.

6. The < 2 mm fraction that passes the screen should be dried (70°C for 48 h), weighed, ground and analyzed for C and N (section 10.3.4); organic material not passing the 2 mm screen, including fine roots (<10 mm diameter) should be similarly processed and analyzed.
7. The mass of stones 2–10 mm in diameter should be recorded.

10.3.3.1 Density and Volume Calculations

The values described below should be calculated for each depth increment sampled from a volumetric pit.

Volume of stones (V_{St}, cm^3) $$V_{St} = M_{St} / D_{St} \tag{10.1}$$

where M_{St}= mass of stones (g), including all stones > 10 mm and 2–10 mm in size
D_{St}= density of stones (g cm^{-3}). This should be determined for a subsample of stones, using water displacement to estimate the volume of the subsample.

Bulk density (D_B, g cm^{-3}). This value represents the average density of all materials occurring in the depth increment.

$$D_B = M_T / V_T \tag{10.2}$$

where M_T= total dry mass (g) of all materials existing in the depth increment (all stones, organic material, soil < 2 mm, and coarse and fine roots).
V_T= total volume (cm^3) of the depth increment

Volume of fine fraction (V_F, cm^3). This value includes the volume occupied by materials passing the 2 mm sieve and the volume of all organic material other than coarse roots.

$$V_F = V_T - V_{St} - V_{CR} \tag{10.3}$$

where V_T= total volume (cm^3) of the depth increment
V_{St}= volume of the stone fraction
V_{CR}= volume of coarse roots (>10 mm diameter), determined from measurement of length and diameter of coarse roots (πd^2 x length).

Fine fraction density (D_F, g cm^{-3}). This value represents the average density of materials passing the 2 mm sieve and all organic material except coarse roots.

$$D_F = M_F / V_F \tag{10.4}$$

where M_F = total dry mass (g) of all materials passing the 2 mm sieve and organic materials not passing the 2 mm sieve or 10 mm sieve, other than coarse roots.
V_F= volume of fine fraction as calculated above.

10.3.4 Elemental Analysis

The C and N concentrations of all soil and organic samples should be determined by dry combustion at high temperature (ca. 1000°C) using elemental analyzers. All samples to be analyzed for C and N should be well homogenized, with a subsample finely ground (40 mesh). This is necessary to ensure samples analyzed are representative of the average C and N concentrations of the whole sample, since the masses used for such analyses are often very small (<0.5–40 mg). Up to 1 kg of the unground sample should also be archived (see section 10.5). Wet combustion methods for C determination, such as Walkley-Black, are not to be used as they can underestimate soil C by 20–30% (Nelson and Sommers 1982).

Some soils may contain significant amounts of inorganic C (carbonates), which must be removed prior to elemental analysis. One method of achieving this is adding small amounts of phosphoric acid to the small sample tins that contain the samples for elemental analysis (Sollins et al. 1999). These samples can then be analyzed after the inorganic C has evolved as CO_2 and the sample has dried. This procedure should not be needed for most samples, but may be necessary for soils from arid and semiarid environments or from lower depths in soils derived from calcareous materials, such as those from glaciated landscapes of the northern states. A subset of samples from locations where carbonates are suspected should analyzed with and without acid treatment to determine if carbonate removal procedures are required for a site and the depth increments to which they should be applied.

Reference soil and plant tissue samples should be included in all sample runs for QA/QC purposes. Available samples from the National Institute of Standards and Technology (NIST) include: Buffalo River Sediment (RM8704) certified at 3.351% C; apple leaves (SRM1515), peach leaves (SRM1547), spinach leaves (SRM1570a), tomato leaves (SRM 1573a) and pine needles (SRM1575a), which have certified or reference values for N of 1.2–5.90%. These plant tissue reference materials do not have certified values for C, but do produce consistent values, which would allow for cross-laboratory comparisons to ensure accuracy in analytical results. The Canadian Centre for Mineral and Energy Technology (CANMET-MMSL) has three certified reference soils (SO-2, SO-3, and SO-4), but these are not certified for C or N. Accurate reference samples for soils can also be prepared from aluminum oxide and EDTA using the method described by Ficklin et al. (2006).

10.3.5 Calculation of Soil C and N Stocks

Carbon stock (S_C, g C m^{-2}). This value is determined for each depth increment sampled. Total C stock for each sample location is obtained by adding the S_C for all depth increments.

$$S_C = T_I \times (10{,}000\ \text{cm}^2/\text{m}^2)[(C_{FS} \times M_{FS}) + (C_{OM} \times M_{OM})] / V_T \qquad (10.5)$$

where T_I= thickness of the depth increment in centimeters
C_{FS}= carbon concentration (g kg^{-1}) of materials passing the 2 mm screen
C_{OM}= carbon concentration (g kg^{-1}) of > 2 mm organic materials (other than coarse roots > 10 mm in diameter)
M_{FS}= mass (kg) of materials passing the 2 mm screen
M_{OM}= mass (kg) of > 2 mm organic materials (other than coarse roots > 10 mm in diameter)
V_T= total volume (cm^3) of the depth increment

The C stocks for the < 2 mm fraction and > 2 mm organic materials can also be calculated separately for each depth increment, if desired. The nitrogen stock (S_N, g N m^{-2}) for each depth increment can be similarly calculated.

C:N ratio. The C:N ratio of fine soil from all depth increments, forest floor materials, and the 2–10 mm soil organic fractions should be determined from the elemental analyses described previously. This ratio serves as a general index of biological availability, and may also be useful as an indicator of long-term changes in soil C stocks or soil C quality in response to global change factors such as elevated atmospheric CO_2 or chronic N deposition.

10.3.6 Problem Soils and Wetlands

The sampling techniques described in this chapter will work well for most upland soils, but soils with certain characteristics may require adjustments to the methodology. In extremely stony soils, larger pits will be required to adequately address stone content and obtain enough fine soil material for representative analysis.

Organic soils may require deeper sampling. They should be sampled deep enough to reach mineral soil horizons that contain relatively little organic matter, even if this requires exceeding the standard sampling depth. Volumetric pits also may not work well for organic soils due to soft material that does not maintain its shape upon sampling, or natural filling of the excavation with water. Sampling in these situations should utilize sharpened volumetric coring devices designed to minimize core compaction, such as those described by Craswell and Castillo (1979), Wardenaar (1987) and Seaby (2001) or reviewed by Sheppard et al. (1993). No single methodology can be recommended for all wetland soils, as they can vary greatly in mineral and organic contents and depth. Methods that work well in one wetland type may not be adequate in another. The technique and tools used in a given location to sample a wetland or other problem soil should be well documented, allowing repetition of the same procedures during future sampling at the location.

In some ecosystems, such as pinyon-juniper woodlands, distinct areas of forest canopy and interspaces will occur. These will have to be sampled using a stratified design that separately measures areas under the canopies and in the interspaces. The percentage of land surface area covered by each stratum will need to be determined.

Certain volcanic rocks can sorb C and N (Sollins et al. 1999). For soils from volcanic sites, a subsample of the stone fraction should be analyzed for concentrations of these elements (Harrison et al. 2003).

Additional considerations also may be needed for other soils. Potential problems and their solutions are discussed in a number of standard soil methodology texts, such as those edited by Carter (1993), Robertson et al. (1999) and Lal et al. (2001).

10.4 Root Biomass

10.4.1 Live Coarse Roots

For mature forest stands, coarse root (>10 mm diameter) biomass can range from 10% to 90% of aboveground woody biomass, but most values range from 20% to 30% (Whittaker and Woodwell 1968, Whittaker et al. 1974, Pastor and Bockheim 1981, Xiao et al. 2003). Therefore, coarse roots can represent a significant amount of the C stored by a forest ecosystem. Destructive methods of sampling coarse structural root biomass are available, but they are labor intensive and require equipment such as fire hoses, dynamite or large winches. Such operations are not feasible at most locations, so we recommend estimating coarse root biomass using published allometric relationships based on diameter at breast height (e.g. Whittaker et al. 1974 and the revised versions of Whittaker's equations in Vadeboncoeur et al. 2007, Omdal et al. 2001, Bond-Lamberty et al. 2002, King et al. 2007, Park et al. 2007). The C concentration of all coarse roots can be assumed to be the same as coarse roots (>10 mm diameter) taken from the volumetric soil pits (see section 10.3.3). If none are encountered, the C concentration of aboveground woody biomass can be used, or samples can be taken from coarse roots leaving the tree base using hole saws or increment corers.

10.4.2 Dead coarse roots

Dead coarse roots could also contain considerable soil C, especially if there have been significant recent losses due to insects, disease, or harvest. To help assess dead coarse root C content, records should be kept of all stumps and standing dead trees during overstory inventory. If the year of tree death (or harvest) is known, it should be recorded. For stumps and dead standing trees, soil near the base should be scraped away from several large structural roots that radiate from the stump, and a decay class should be recorded, using codes similar to the log decay classes of Triska and Cromack (1979):

1 = recently dead, both outer and inner bark intact

2 = root is firm, with some inner and possibly outer bark remaining, may show partial invertebrate and/or fungal colonization into sapwood
3 = generally associated with extensive invertebrate and fungal colonization and color change in sapwood and heartwood. Some fragmentation has occurred, but shape of the root and stump has not yet begun to change.
4 = roots and stump completely colonized by invertebrates and fungi. They are highly fragmented and roots have begun to collapse.
5 = roots are highly fragmented, highly collapsed, and generally have sunken into the surrounding soil layers.

The dead coarse root data will enable calculation of the C involved, by multiplying dead coarse root mass by C concentration. Mass can be calculated by using the allometric equations described above and adjusting results for decay class and/or time since death. At present, researchers will need to obtain the most appropriate data they can find for dead coarse root decay rates and densities to use in adjusting allometric results. As the literature on coarse root decay becomes more extensive, the accuracy of such adjustments will improve and previous estimates can be recalculated to improve their accuracy. A record of current stumps and decay classes will make such back calculation possible in the future.

10.4.3 Fine Roots

Fine roots (<10 mm in diameter) are not separated from the > 2 mm organic material during the sieving procedure, since the roots obtained would not represent the majority of fine roots present in a location. This is the result of many fine root fragments passing through the 2 mm screen and the existence of substantial fine root biomass in some forest floor horizons. It should be noted that the C content of all fine roots will be accounted for during analyses of samples from the forest floor, fine soil fraction, and organic material not passing the 2 mm screen.

The C stock of fine root biomass is often much lower than that found in the forest overstory or soil pools (Xiao et al. 2003, also compare MacDonald et al. 1991, Reed et al. 1994, and Burton et al. 2004). However, it is a very dynamic soil C fraction and changes in it may be important in explaining year-to-year variability in the C exchange at a location. As a result, it may be desirable to have an estimate of this C pool. In such cases, fine roots can be hand sorted from subsamples of the existing forest floor, fine soil fraction and > 2 mm organic material samples or from separate samples collected specifically for assessing fine roots. At some locations, separate samples collected using cores (5–10 cm diameter) or smaller diameter pits may provide the most efficient method of obtaining fine root samples through the first 20 cm of mineral soil.

The fine root fraction should be removed by hand sorting and further divided into smaller size classes (i.e. < 0.5, 0.5–1, 1–2, and 2–10 mm) more representative of root functional roles, and washed free of all adhering soil or organic matter (Burton et al. 2004). Dead and live fine roots can be separated during the sorting procedure based on physical appearance and structural integrity (Burton et al. 2004). To fully capture fine roots, soil and forest floor materials from which fine roots have been hand sorted should then be elutriated (separated using an air/water flotation system, Smucker et al. 1982). Elutriation can increase recovery of very fine roots (< 1 mm) by 30–40% or more relative to dry sieving and hand sorting (Pregitzer et al. 2008). Sorting fine roots and processing elutriated samples are very labor intensive and time consuming, which should be considered when determining whether to include fine roots in a sampling scheme.

10.5 Archiving Samples

All forest floor, mineral soil and coarse root samples should be archived. Archived samples could be used at future times to examine a variety of factors other than total C stocks, such as exchangeable base cations, P content, or soil C fractions. The ability of today's researchers to examine effects of management practices and global change on slowly changing soil factors is often limited by the lack of archived soil samples from earlier sampling efforts. To maximize the knowledge that can be gained from the samples collected using the methods described in this chapter, all unused portions of the samples ground for C and N analysis should be archived. Unground samples of up to 1 kg should also be archived, as they may be preferred for some future analyses. Dry, archived samples should be kept in inert containers with tight fitting lids and permanent labels (Boone et al. 1999). Glass and plastic are least likely to react with the soil and forest floor materials. Archived samples should be stored in dry rooms, which preferably do not undergo large temperature fluctuations. Boone et al. (1999) provide a useful discussion of additional items to consider when archiving soil samples.

Literature Cited

Bond-Lamberty B, Wang C, Gower ST (2002) Aboveground and belowground biomass and sapwood area allometric equations for six boreal tree species of northern Manitoba. Canadian Journal of Forest Research 32: 1441–1450

Boone, RD, Grigal DF, Sollins P, Aherns RJ, Armstrong DE (1999) Soil sampling, preparation, archiving and quality control. In: Robertson GP, Coleman DC, Bledsoe CS, Sollins P (Eds.). Standard soil methods for long-term ecological research. Oxford University Press, New York, pp. 3–28

Brye KR, Morriss TL, Miller DM, Formica SJ, Van Eps MA (2004) Estimating bulk density in vertically exposed stoney alluvium using a modified excavation method. Journal of Environmental Quality 33: 1937–1942

Burton AJ, Pregitzer KS, Crawford JN, Zogg GP, Zak DR (2004) Simulated chronic NO_3^- deposition reduces soil respiration in northern hardwood forests. Global Change Biology 10: 1080–1091

Carter MR (Ed.) (1993) Soil sampling and methods of analysis. Lewis Publishers, Boca Raton, FL.

Craswell ET, Castillo EG (1979) A convenient device for taking large, undisturbed samples of paddy soil. International Rice Research Newsletter 4: 22

Ficklin RL, Dresbach RI, Richardson JD (2006) A reliable, inexpensive method for creating samples of known carbon and nitrogen concentrations for use in dry combustion C-N analyzers. American Laboratory News 38(17): 26–27

Harrison RB, Adams AB, Licata C, Flaming B, Wagoner GL, Carpenter P, Vance ED (2003) Quantifying deep-soil and coarse-soil fractions: Avoiding sampling bias. Soil Science Society of America Journal 67: 1602–1606

Howard RF, Singer MJ (1981) Measuring forest soil bulk density using irregular hole, paraffin clod, and air permeability. Forest Science 27: 316–322

King JS, Giardina C, Pregitzer KS, Friend AL (2007) Biomass partitioning in red pine (*Pinus resinosa*) along a chronosequence in the Upper Peninsula of Michigan. Canadian Journal of Forest Research 37: 93–102

Lal R, Kimble JM, Follett RF, Steward BA (Eds.) (2001) Assessment methods for soil carbon. Lewis Publishers, Boca Raton, FL

MacDonald NW, Burton AJ, Jurgensen MF, McLaughlin JW, Mroz GD (1991) Variation in forest soil properties along a Great Lakes air pollution gradient. Soil Science Society of America Journal 55: 1709–1715

Muller RN, Hamilton ME (1992) A simple, effective method for determining the bulk density of stony soils. Communications in Soil Science and Plant Analysis 23: 313–319

Nelson DW, Sommers LE (1982) Total carbon, organic carbon, and organic matter. *In* Methods of Soil Analysis, Part 2. Chemical and Microbiological Properties. Page AL, Miller RH, Kenney DR (Eds.) ASA Monograph 9. American Society of Agronomy - Soil Science Society of America, Madison, WI, pp. 539–579

Omdal DW, Jacobi WR, Shaw CG III (2001) Estimating large-root biomass from breast-height diameters for ponderosa pine in northern New Mexico. Western Journal of Applied Forestry 16: 18–21

Page-Dumroese DS, Jurgensen MF, Brown RE, Mroz GD (1999) Comparison of methods for determining bulk densities of rocky forest soils. Soil Science Society of America Journa*l* 63: 379–383

Park BR, Yanai RD, Vadeboncoeur MA, Hamburg SP (2007) Estimating root biomass in rocky soils using pits, cores, and allometric equations. Soil Science Society of America Journal 71: 206–213

Pastor J, Bockheim JG (1981) Biomass and production of an aspen - mixed hardwood - spodosol ecosystem in northern Wisconsin. Canadian Journal of Forest Research 11: 132–138

Pregitzer KS, Burton AJ, King JS, Zak DR (2008) Soil respiration, root biomass, and root turnover following long-term exposure of northern forests to elevated atmospheric CO_2 and tropospheric O_3. New Phytologist *in press* (doi:10.1111/j1469-8137.2008.02564.x)

Reed DD, Pregitzer KS, Liechty HO, Burton AJ, Mroz GD (1994) Productivity and growth efficiency in sugar maple forests. Forest Ecology and Management 70: 319–327

Robertson GP, Coleman DC, Bledsoe CS, Sollins P (Eds.) (1999) Standard soil methods for long-term ecological research. Oxford University Press, New York

Seaby D (2001) Designs for one-man, two-stage, samplers for obtaining undisturbed cores of peat over 1 m long. Forestry 74: 79–83

Sheppard MI, Tarnocai C, Thibault DH (1993) Sampling organic soils. *In.* Carter MR, (Ed.). Soil Sampling and Methods of Analysis. Lewis Publishers, Boca Raton, FL

Sollins P, Glassman C, Paul EA, Swanston C, Lajtha K, Heil JW, Elliott ET (1999) Soil carbon and nitrogen: pools and fractions. Pages 89–105. *In* Robertson GP, Coleman DC, Bledsoe CS, Sollins P (Eds.). Standard soil methods for long-term ecological research. Oxford University Press, New York

Smucker AJM, McBurney SL, Srivanstava AK (1982) Quantitative separation of roots from compacted soil profiles by the hydropneumatic elutriation system. Agronomy Journal 74: 500–503

Triska FJ, Cromack K (1979) The role of wood debris in forests and streams. *In* Forests: Fresh perspectives from ecosystem analysis. Proceedings, 40th Biology Colloquium. Oregon State University, Corvallis, pp. 171–191

Vadeboncoeur MA, Hamburg SP, Yanai RD (2007) Validation and refinement of allometric equations for roots of northern hardwoods. Canadian Journal of Forest Research 37: 1777–1783

Wardenaar ECP (1987) New hand tool for cutting peat profiles. Canadian Journal of Botany 65: 1772–1773

Whittaker RH, Woodwell GM (1968) Dimension and production relations of trees and shrubs in the Brookhaven Forest, New York. Journal of Ecology 56: 1–25

Whittaker RH, Bormann FH, Likens GE, Siccama TG (1974) The Hubbard Brook ecosystem study: Forest biomass and production. Ecological Monographs 44: 233–252

Xiao CW, Yuste JC, Janssens IA, Roskams P, Nachtergale L, Carrara A, Sanchez BY, Ceulemans R (2003) Above- and belowground biomass and net primary production in a 73-year-old Scots pine forest. Tree Physiology 23: 505–516

Chapter 11
Quantifying Soil Respiration at Landscape Scales

John B. Bradford and Michael G. Ryan

Abstract Soil CO_2 efflux, or soil respiration, represents a substantial component of carbon cycling in terrestrial ecosystems. Consequently, quantifying soil respiration over large areas and long time periods is an increasingly important goal. However, soil respiration rates vary dramatically in space and time in response to both environmental conditions and biological activity. Our objective in this chapter is to characterize the challenges in capturing this variability and accurately estimating soil respiration. We first review approaches to collecting individual soil respiration measurements, with particular focus on their applicability to landscape-scale studies. We then identify the major sources of variability in respiration rates and discuss how individual measurements can be structured in space and time to capture that variability. Lastly, we present a set of recommendations for an integrated approach that combines spatially distributed measurements with temporally intensive measurements to develop annual, landscape-scale soil respiration estimates.

Keywords Net ecosystem carbon balance, soil carbon dioxide efflux, spatial and temporal scaling

---

J.B. Bradford
US Forest Service, Northern Research Station, 1831 Hwy 169 E., Grand Rapids MN 55744.
E-mail: jbbradford@fs.fed.us

M.G. Ryan
US Forest Service Rocky Mountain Research Station, 240 W. Prospect Ave., Fort Collins, CO 80526, Affiliate Faculty, Graduate Degree Program in Ecology, Colorado State University, Fort Collins, CO
E-mail: mgryan@fs.fed.us

C.M. Hoover (ed.) *Field Measurements for Forest Carbon Monitoring*,

11.1 Introduction

Soil surface CO_2 efflux is the release of carbon dioxide from the soil surface to the atmosphere and is commonly called soil respiration. This flux comprises 50–80% of ecosystem respiration (Davidson et al. 2002a, Giardina and Ryan 2002) and consists of respiration from roots and associated mycorrhizae and from heterotrophic microbes using root exudates and recent and older organic material as an energy substrate (Wiant 1967a, Anderson 1973). Instantaneous CO_2 flux rates range from near zero during winter to >10 μmol m^{-2} s^{-1} for high productivity ecosystems during the growing season (Raich et al. 2002) and annual estimates range from less than 200 g C $m^{-2}year^{-1}$ in xeric systems to nearly 2000 g C $m^{-2}year^{-1}$ in wet temperate forests (Hibbard et al. 2005).

As with many ecological processes, interest in soil respiration has shifted from addressing site-specific or treatment-related questions to characterizing respiration rates for large areas over long time periods (Underwood et al. 2005). Large-area and long-term estimates of soil respiration are needed to: (1) reduce uncertainties in landscape, regional and global carbon budgets (Law et al. 2002), (2) characterize the spatial and temporal dynamics in plant physiological processes, including belowground carbon allocation (Giardina and Ryan 2002), (3) facilitate direct comparisons with eddy-covariance measurements (Pypker and Fredeen 2002, Kutsch et al. 2005, Tang and Baldocchi 2005, Tang et al. 2005a), and (4) provide parameterization and validation for ecological simulation models (Chen et al. 2000, Soegaard et al. 2000, Tate et al. 2000). There is also a need to improve understanding of mechanisms controlling soil CO_2 fluxes through experimentation to advance models that provide continuous estimates of fluxes and processes contributing to net ecosystem exchange of CO_2.

Large-area and long-term estimates of soil respiration are complicated by the high variability of soil respiration in both space and time and by the limited spatial and temporal extent of actual measurements. Soil respiration has been shown to vary dramatically in temporal scales ranging from hours (Ekblad et al. 2005) to years (Raich et al. 2002) and in spatial scales ranging from meters (Tang and Baldocchi 2005) to regions (Reichstein et al. 2003). In addition, individual soil respiration measurements typically cover less than 0.25 m^2 and represent only a snapshot of a few minutes (Lavigne et al. 1997, Murthy et al. 2003). These two realities complicate the process of generating accurate large-area and long-term soil respiration estimates because, unlike many ecological processes, soil respiration has not been clearly linked to aboveground structural or functional patterns (Fahey et al. 2005) that are easily mapped with remote sensing (although see Reichstein et al. 2003, Tang et al. 2005a). Studies are beginning to explicitly characterize the scales and drivers of this spatial and temporal variability and these results will undoubtedly contribute to the up-scaling of soil respiration.

Our objectives in this chapter are: (1) to briefly describe the methods for measuring soil respiration, focusing on the applicability of these methods to generating landscape-level annual estimates, (2) to identify the sources of variability in soil respiration and characterize approaches to scaling soil respiration over space and time and (3) to recommend standard methods for quantifying annual, landscape-level soil respiration fluxes. Our overall goal is to address the question: how do we obtain large-scale long-term estimates of a flux that can only be measured for very small areas over very short intervals? Several detailed reviews have examined small-scale soil respiration methods and controls over soil respiration (Hanson et al. 2000, Rustad et al. 2000, Davidson et al. 2002b, Hibbard et al. 2005, Ryan and Law 2005).

Soil respiration has been recognized as a primary component of ecosystem carbon dynamics for several decades (e.g. Lundegardh 1927, Witkamp 1966, Schulze 1967, Wiant 1967b, c, Reiners 1968). Initial measurements of soil respiration provided insight into relative rates between locations and through time, but were not able to accurately quantify absolute rates. As interest in quantifying absolute rates of soil respiration grows, researchers are becoming increasingly critical of measurement techniques. Laboratory (Nay et al. 1994, Widen and Lindroth 2003, Butnor and Johnsen 2004) and field (Le Dantec et al. 1999, Janssens et al. 2000, Pumpanen et al. 2003) examinations have led to modifications of existing approaches and entirely new techniques to measure soil respiration. Three general methods for quantifying soil respiration are currently in use: chambers using a closed system (dynamic or static), chambers using an open system, and flux gradient sensors. In addition to these techniques, some approaches have arisen for quantifying soil respiration under snow.

11.2 Chambers Using a Closed System

11.2.1 Approach

Closed chamber systems for measuring soil respiration are currently the most common and represent the only commercially available systems. Closed systems estimate flux by measuring change in CO_2 concentration inside a closed chamber over the soil surface, usually fixed onto a plastic ring embedded into the soil. These systems are named 'closed' because no air is exchanged between the chamber and the outside environment during measurement. However, between measurements the system is open to the environment. Most closed systems utilize a dynamic approach that continually circulates air from the chamber to an infrared gas analyzer and back to the chamber (Norman et al. 1992). Other systems avoid circulating air and use a static approach that measures CO_2 in the chamber by extracting and analyzing gas in a syringe (Parkinson 1981), absorbing CO_2 in soda lime within the chamber (Edwards 1992), or, in the future, using laser spectroscopy (Gianfrani et al. 2004) or small infrared gas analyzers (for example, Vaisala CARBOCAP® Carbon Dioxide Probe GMP343, Vaisala Group, Vantaa, Finland) inside the chamber to continuously monitor CO_2 concentration. Static systems have been demonstrated to underestimate high fluxes and overestimate low fluxes (Nay et al. 1994, Pongracic et al. 1997, King and Harrison 2002), perhaps because of problems with the rate of absorbtion of CO_2 onto the soda lime, or because of high CO_2 concentrations inside the chamber impeding diffusion.

Regardless of the approach, all closed systems quantify the rate of increase in CO_2 concentration (µmolCO_2 mol^{-1}air s^{-1}) inside a chamber of known volume. This rate is divided by the volume (mol) of air in the chamber to yield flux in CO_2 per unit time (µmolCO_2 s^{-1}), and is divided by the surface area covered by the chamber to estimate temporal CO_2 flux per area (µmolCO_2 m^{-2} s^{-1}).

11.2.2 Challenges

Two challenges are inherent in closed systems. First is that the estimated soil respiration depends on the total volume of the chamber, tubing, IRGA and soil pore space (in moles of air). Soil pore space may influence the calculated efflux rate by serving as additional volume where CO_2 concentration will increase during measurement, effectively increasing the total system volume and therefore decreasing estimated flux rates. Consequently, soils with extremely high soil pore space will require more CO_2 efflux to yield the same change in CO_2 concentration, which is the measured indicator of efflux rate in closed systems. Although some studies use a nominal system volume calculated from the above-soil collar and chamber volume, plus tubing and IRGA gas path, effluxed CO_2 is also stored in the soil pore space. Rayment (2000) estimated soil pore space by combining CO_2 efflux rate and initial rate of change in CO_2 concentration in an open system. Results from this study indicate that the equivalent depth of air in the soil averaged 15.5 mm, which translates into underestimation of soil respiration of 9.1% if soil pore space were ignored in closed systems.

Another approach to quantifying the total system volume is to add CO_2 at a constant known rate to the closed system during measurement and compare the calculated flux from a paired measurement without this standard addition (M.G. Ryan personal communication). In this "standard addition" approach, the total system volume (mol air) is calculated by dividing the constant CO_2 addition (µmolCO_2 s^{-1}) by the amount that this addition increases the rate of change in CO_2 concentration (µmolCO_2 mol^{-1}air s^{-1}). This increased rate of change is simply the rate of change with standard addition minus the rate of change without standard addition. Regardless of the approach, closed systems require accurate representation of total system volume. The magnitude of pore space depends on soil properties, including moisture, texture and bulk density, which can vary through time and between sites. Thus, the importance of quantifying total system volume will depend on the ecosystem; consistently wet areas with fine textured soils may not require quantifying pore space whereas locations with coarse textured soils and high seasonal variation in soil moisture likely require multiple measures of pore space in each year (Butnor and Johnsen 2004, Butnor et al. 2005). Measuring the volume of each measuring point will yield more precise flux estimates than assuming a standard volume across the site.

The second challenge for closed systems is the possibility of pressure differences between inside and outside the chamber influencing perceived CO_2 flux rates. Soil pore space has very high CO_2 concentrations, which represent a large reservoir of CO_2. High pressures outside the chamber, caused by variable wind speeds, can force CO_2-rich air from the soil pore space into the chamber, increasing CO_2 concentration and artificially elevating estimates of soil respiration (Davidson et al. 2002b, Bain et al. 2005). As a consequence, most closed systems have been equipped with vents to equalize pressure; however, some vented systems have been shown to underestimate soil respiration, possibly as a result of leaking CO_2 through the vent (Conen and Smith 1998).

11.3 Chambers Using an Open System

11.3.1 Approach

Chambers using the open system estimate flux by precisely measuring the rate of airflow through the chamber and the inlet and outlet CO_2 concentrations at equilibrium (Fang and Moncrieff 1996, Liang et al. 2004, Butnor et al. 2005). These are called 'open' systems because air is exchanged between the outside and the chamber. Open systems that are operated for continuous measurements (for example, Palmroth et al. 2005) typically have the chamber closed during the entire measurement period. Problems with changing the environment during long-term measurements are managed by alternating the chamber between two adjacent collars every two days.

When open chambers are initially placed over the soil surface, this difference is zero and as CO_2 builds up in the chamber, the difference increases until it reaches a steady state at which CO_2 leaving the chamber is in equilibrium with CO_2 efflux from the soil. At this point, the difference in CO_2 concentration (μmolCO_2 mol^{-1}air) between air in and out of the chamber can be multiplied by the flow rate (mol air s^{-1}) and divided by the soil surface area covered by the chamber to calculate the soil respiration rate (μmolCO_2 m^{-2} s^{-1}). Open chambers require a reservoir of input gas of consistent CO_2 concentration in order to avoid fluctuations in the reference CO_2 concentration that will increase measurement variability.

11.3.2 Challenges

One potential challenge associated with the open chamber approach is the possibility of the elevated CO_2 concentrations in the chamber inhibiting CO_2 efflux from the soil. While this inhibition could also occur in closed systems, the potential for bias in open systems is greater because the CO_2 concentration inside the chamber is elevated for the duration of the measurement, whereas it is only elevated at the end of the measurement in the closed system. Observational evidence for elevated CO_2 concentration negatively impacting efflux is limited (Amthor 2000), but theoretically, elevated chamber CO_2 concentrations could decrease microbial and root metabolic activity and/or slow CO_2 diffusion from the soil by decreasing the gradient between CO_2 concentrations in soil pore space versus chamber air space. The actual consequence may be a combination of these two effects, decreasing CO_2 diffusion from lower layers. The primary obstacle to the open chamber approach is the practical difficulty of using it to acquire enough measurements to characterize a landscape. The instruments and air reservoir utilized by the open chamber method are more difficult to transport than the closed chamber apparatus. In addition, the open chamber approach requires several minutes to obtain a measurement, which represents substantial time investment. The combination of these two practical limitations makes open chambers difficult to apply to landscape-scale studies. However, the open chamber approach does have the advantage that it maintains relatively constant and ambient CO_2 concentration inside the chamber, which makes it well suited to collecting continuous measurements (discussed further below).

11.4 Soil CO_2 Gradient

11.4.1 Approach

An alternative to chamber methods for measuring soil respiration is to measure CO_2 concentration at multiple depths in the soil profile and use this gradient along with the CO_2 diffusivity to model soil CO_2 efflux (de Jong and Schapper 1972, Wagner and Buyanovsky 1983, Burton and Beauchamp 1994, Tang et al. 2003, Jassal et al. 2005). Quantifying the vertical gradient in soil CO_2 concentration is accomplished by burying small infrared detectors in the soil with openings to the soil pore space at specified depths. CO_2 diffusivity is calculated from soil properties and CO_2 diffusion coefficient in air (measured empirically for reference conditions and corrected for on-site conditions) (Tang et al. 2003).

11.4.2 Challenges

Although the CO_2 gradient approach avoids many of the challenges inherent in chamber measurements, it is limited by the fine spatial extent of its measurements, the difficulty of collecting multiple measurements across a landscape, and the potentially high variability of CO_2 diffusivity (Tang et al. 2003). The set of sensors in the CO_2 gradient method sample only a very small area of soil surface, requiring many more measurements to accurately characterize a landscape, especially in ecosystems with high spatial variability at small scales. In addition, the sensors for the CO_2 gradient approach must be buried prior to measurement (to avoid disturbing the soil CO_2 concentrations and to allow CO_2 concentrations inside the sensors to reach equilibrium with the soil.) Consequently, individual sets of sensors are required for each sampled location, making large sample sizes financially unfeasible. Lastly, the estimates for soil respiration from the CO_2 gradient approach utilize CO_2 diffusivity, which depends on soil and air conditions that can both vary in space and time. Accurately characterizing CO_2 diffusivity at multiple locations over several time periods could dramatically increase the difficulty of using the CO_2 gradient approach to estimate landscape-scale soil respiration.

11.5 Under-Snow Measurements

11.5.1 Approach

Many ecosystems are characterized by long, cold winters and sufficient snowfall to create snowpack that persists for many months (Sommerfeld et al. 1993). Soil conditions under snow are highly variable, but can frequently include temperatures above freezing and/or high moisture availability, creating an environment suitable for respiration (Brooks et al. 1996, Brooks et al. 1997). Consequently, quantifying annual soil respiration in snowy locations requires measurement of soil CO_2 efflux under the snow. One technique for quantifying under-snow soil respiration is to use the difference between the CO_2 concentration above and below the snow and properties of the snowpack to model CO_2 efflux through the snow (Hubbard et al. 2005). Soil surface CO_2 concentrations are measured by inserting a probe through the snowpack to the soil surface. The probe is open to the air at the bottom, and contains tubing that is connected to a backpack gas analyzer and pump. Using Fick's first law, flux can be calculated from snowpack depth, porosity, and temperature and CO_2 molecular density and diffusion in air and tortuosity (Massman et al. 1995).

11.5.2 Challenges

Individual measurements with this under-snow method are relatively rapid, requiring less than a minute each, which facilitates the collection of numerous points across large areas. However, this technique requires enough snowpack to create a substantial gradient in CO_2 concentration between the soil surface and the snow surface, meaning that it is only feasible for locations with substantial snowfall that creates consistent snowpack across the landscape and throughout the winter. In addition, this approach could be sensitive to snowpack compaction caused by the person conducting the measurements, which could either increase perceived respiration by increasing the barrier to CO_2 diffusion, or decrease the perceived respiration rate by decreasing the barrier. Some studies have found that under-snow respiration rates are highly susceptible to pressure pumping from varying wind speeds (Massman et al. 1997, Takagi et al. 2005). This suggests the need for many measurements, which is especially true in this method where each measurement represents only a very small point of soil surface. Lastly, this method is relatively new, and at least one study has suggested that it may underestimate soil CO_2 efflux (Takagi et al. 2005).

11.6 Generating Landscape and Annual Estimates

11.6.1 Sources of Variability in Soil Respiration

Variability in soil respiration can be conceptualized as temporal variation, which represents differences through time at individual locations, and spatial variation, which represents differences between locations (Fig. 11.1). Spatial variation in soil respiration occurs at scales as small as a meter (Murthy et al. 2003), where topography and vegetation patch structure influence microclimate, to intermediate scales where soil properties and ecosystem type impact carbon substrate and root density/activity (King et al. 2001, Sulzman et al. 2005), to large scales where climatic conditions dictate overall conditions (Campbell and Law 2005). Likewise, temporal variation in soil respiration ranges from high frequency, short-term variations in wind speed that impact pressure pumping (Massman et al. 1997) through intermediate scales of hours to days where pulse precipitation events and diurnal temperature

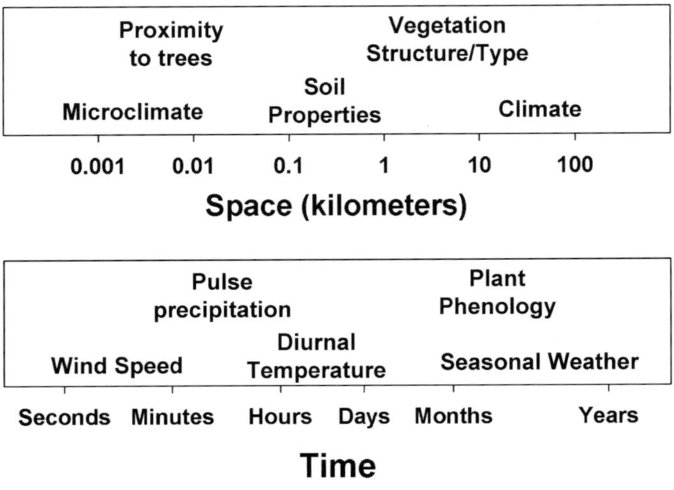

Fig. 11.1 Sources of variability in soil respiration across space and time

fluctuations influence decomposition rates and plant activity (Kabwe et al. 2005, Tang et al. 2005b) to seasonal weather and plant phenological variations that occur across months to years (Chambers et al. 2004, Hubbard et al. 2005, Sulzman et al. 2005).

11.6.2 Approaches to Extrapolating Soil Respiration

Soil respiration varies through time and space in response to soil temperature, soil moisture and vegetation composition. In many forest systems where soils rarely become extremely dry, soil temperature alone is a relatively good predictor of soil respiration rates and has been frequently used to estimate soil respiration (Rodeghiero and Cescatti 2005). In arid and semi-arid ecosystems, where soil moisture can decrease to levels that limit microbial and plant activity, variation in soil moisture must be considered to accurately estimate annual soil respiration (Chambers et al. 2004, Xu et al. 2004). Plant activity also influences soil respiration by dictating diurnal and seasonal trends in root respiration (Wang et al. 2005) and by influencing spatial and temporal patterns of microclimatic soil temperature and moisture status (Palmroth et al. 2005). Generating landscape-level annual or multi-year estimates of soil respiration requires knowing what influences soil respiration and at what scales those drivers fluctuate.

11.6.2.1 Temporal Scaling

Previous studies have identified three general temporal scales at which variation in soil respiration occurs: seasonal fluctuations due to climate and plant phenology (Rayment and Jarvis 2000), diurnal patterns controlled by temperature and plant activity (Tang et al. 2005a), and episodic peaks lasting for hours to days that are driven by pulse weather events (Reth et al. 2005). These results suggest that soil respiration measurements should include at least some measurements at each of these scales.

Some studies have generated annual estimates of soil respiration by using a combination of individual monthly soil respiration measurements (one measure per month) along with a few isolated diurnal measurements of soil respiration (at least two measures per day) (Tang and Baldocchi 2005). The monthly measurements provide insight into the seasonal variation and the diurnal measurements quantify fluctuations within individual days. These measures are typically incorporated into a simple statistical model for estimating soil respiration based on soil temperature (Rayment and Jarvis 2000, Zheng et al. 2005a) and occasionally soil moisture (Chambers et al. 2004, Martin and Bolstad 2005), possibly with separate functions for diurnal vs. seasonal fluctuations in these drivers (Litton et al. 2004). This approach has the advantage of being relatively easy to implement in the field, but relies on the assumption that the monthly measures are frequent enough to capture seasonal trends, and on the assumption that diurnal patterns do not dramatically change throughout the year. In addition, this approach ignores the potentially important effects of pulse weather events, which have been shown to influence soil respiration in many ecosystems (Xu et al. 2004).

Other studies have utilized near-continuous measurements for all or part of the year to characterize temporal patterns of soil respiration (Liang et al. 2003, Butnor et al. 2005). This technique requires an automated system for measuring soil respiration, which can be applied to the chamber approach or the CO_2 concentration gradient approach. For the chamber approach, an automated system can maintain multiple chambers and requires hardware for closing and opening the chambers before and after the measurements, as well as a pump, gas analyzer and control system for dictating gas flow and data storage. This automated system cycles through the chambers collecting measurements for each chamber approximately every 1–1.5 hours. Automated systems utilizing closed systems can estimate total chamber volume (and thus overcome the primary challenge of closed chambers discussed above) by adding a standard addition of CO_2 once daily. Continuous measurements from the CO_2 concentration gradient approach are simpler, requiring only multiple buried detectors and a control system for data storage (Tang et al. 2003), but they incorporate the challenge of limited spatial extent, as mentioned above. Regardless of the approach, continuous or near-continuous measurements provide detailed insight into the temporal dynamics of soil respiration and the relationship between respiration rates and driving variables at all scales, including unpredictable pulse events. The only disadvantage of continuous measurements is the expense and operational time required to establish and maintain the automated system. A commercial automated soil respiration measurement sysytem is now available to simplify capturing information on temporal variability (LICOR 8100, LICOR, Inc, Lincoln, NE, USA).

11.6.2.2 Spatial Scaling

Previous studies have examined the effect of spatial patterns of soil temperature, soil moisture and vegetation composition on soil respiration and used these relationships to estimate soil respiration for areas ranging from plots to the globe. At the smallest scale, proximity to trees has been shown to influence soil respiration rates through root respiration and by intercepting precipitation (Tang and Baldocchi 2005). Other studies have chosen to account for this variability by examining overall variation within individual plots and quantifying the number of soil respiration measurements necessary to accurately characterize the plot. Estimates range from 25–30 points/ha in forest plantations to 40–50 points/ha in natural forests (Davidson et al. 2002b, Yim et al. 2003, Adachi et al. 2005) to generate an estimate with a standard error that is within 10% of the mean. The number of samples required may be lower under snowpack (Hubbard et al. 2005) when plant activity is minimal and temperature/moisture conditions are more spatially homogeneous. The sample size required to characterize a given area could likely be decreased by identifying controls over variability at smaller scales and quantifying the distribution of those controls within individual plots, and by using larger collars (such as the 200 mm collar for the LICOR 8100, or even 250 mm collars).

At slightly larger spatial scales soil respiration is influenced by tree density, live biomass, species composition and vegetation type (Litton et al. 2003, Bolstad and Vose 2005, Campbell and Law 2005, Zheng et al. 2005b) and soil properties (Dilustro et al. 2005). Although many studies have observed these controls, few studies have attempted to directly scale ground measurements to estimate soil respiration for areas as large as entire watersheds. One exception is Fahey et al. (2005) who estimated soil respiration for the Hubbard Brook Experimental Forest using spatially dispersed periodic measurements across the watershed and found strong relationships between temperature and respiration rate but no clear relationships between forest composition and respiration. Thus, Fahey et al. (2005) quantified the relationship between respiration and soil temperature and estimated annual flux from continuous measurements of soil temperature. Large-scale estimates of soil respiration have been modeled from relationships with air temperature, precipitation and vegetative productivity (Aikio et al. 2000, Raich et al. 2002, Reichstein et al. 2003). However, there is some evidence that relationships with temperatures from the air or soil surface alone may not adequately capture the temperature dynamics that influence soil respiration (Reichstein et al. 2005).

11.7 Recommendations

Our recommendations include both a suggested protocol for collecting individual soil respiration measurements and a strategy for structuring these measurements in space and time to generate landscape-level annual soil respiration budgets.

11.7.1 Protocol for Individual Measurements

We propose using permanent soil collars and a closed system gas exchange measurement, with a measurement protocol similar to that used by the current LiCOR LI-6400 measurement system and soil chamber. The advantages of this approach are (1) the large chambers sample 6x the area of the standard LiCOR chamber and reduce within-plot variability (from a CV of ~100% to ~25% in a recent study; Ryan et al. unpublished data); (2) fixed chamber locations allow separation of environmental variability from spatial heterogeneity; (3) simple, quick measurements enable rapid sampling for spatial heterogeneity and allow the detection of flux differences among treatments or different vegetation conditions; (4) the closed system measurement is reliable for many soils (Butnor et al. 2006), although it tends to underestimate fluxes from soils that are especially porous; and (5) scrubbing the CO_2 to below ambient levels prior to measurement and measuring CO_2 through ambient levels has been shown to reduce bias from the accumulation of CO_2 in the chamber headspace. Disadvantages of the method are that measurements and models need to be developed to extrapolate between the point measurements in time, installation of collars may damage roots and time is needed for recovery, and the closed system approach can underestimate fluxes for some conditions.

11.7.1.1 Materials

Chamber collars are made from 10″ (25 cm) inside diameter PVC sewer pipe with a bevel on the end to be inserted into the litter. Collar height is designed so that the lower part of the collar contacts a dense portion of the soil surface to minimize advective air flows and the upper portion is 5 cm above the litter surface. Typically, the collar is inserted through the litter and organic layer until the bottom contacts the mineral soil. We use cheap serrated knives to cut the litter around the collar and then insert the collar through the cut slot. We use a rubber mallet and a short piece of 2 x 4 on top of the collar to seat it into the mineral soil. Measurements are made using a homemade gas analysis system, with control provided by a Campbell data logger (23x or 10x) and the LiCOR 820 gas analyzer. The end cap is designed to fit the chamber collar. While we have not yet tested it, we believe that the LICOR 8100 with 200 mm diameter collars would be suitable for these measurements and the large collar should reduce measurement effort.

11.7.1.2 Measurement Protocol

The closed system estimates flux by measuring the rate of increase in CO_2 concentration (μmol mol^{-1} s^{-1}). Flux is calculated by using the standard addition protocol (outlined above) and should be completed often enough to capture temporal dynamics in soil water status, which can influence soil pore space. In the absence of standard addition volume estimates, system volume can be roughly estimated by summing the volume of component parts and adding an estimate of soil pore space. Temperature, pressure and collar dimensions must be known to calculate the molar volume of air in the chamber. Nominal volume is calculated using the volume of the chamber plus tubing plus LiCOR gas path (a constant) plus the volume of the collar above the litter (varies from collar to collar).

Face the chamber in the direction of the prevailing wind and measure ambient (air) CO_2 concentration. CO_2 concentrations in the chamber should center around this value, but in practice, this is difficult to exactly achieve given variability in flux rates. Choose values for the lower bound for CO_2 concentration during the scrub and the delay time to account for scrub overshoot that result in the measurement being taken while the internal CO_2 concentration crosses ambient. Higher fluxes require lower scrub values and shorter delay times. While LiCOR recommends averaging 2–3 readings per collar, we have found this variability trivial compared to the variability among collars (M.G. Ryan unpublished data); so we take one measurement per collar per sampling trip. Soil (and perhaps litter) moisture and soil temperature are also measured for each chamber at depth(s) representative of soil carbon and rooting depth intensities.

11.7.2 Spatial and Temporal Sampling Strategy

Scaling soil CO_2 efflux measurements from instantaneous (or very short-term) measurements of small areas to landscape-scale annual soil respiration budgets requires addressing all of the sources of spatial and temporal variability. Partitioning variability into spatial and temporal components provides a starting point for understanding and managing this variation. Controls over spatial variability, like micro-environmental conditions, soil properties, ecosystem type and climate (not weather), must be spatially mapped across the landscape but need not be examined at multiple times throughout the year because these spatial controls may change over several years, but will be unlikely to change within individual years. However, controls over spatial variability can influence the temporal pattern of soil respiration (e.g. seasonal root respiration may differ between forests and meadows). Therefore, temporal variability must be characterized within each general category of spatial driver at a resolution that captures important seasonal, daily and possibly hourly fluctuations. Once this temporal variability is quantified, annual budgets can be generated for each spatial category and summed to yield landscape-scale estimates. To characterize both the spatial and temporal variability in soil respiration, we propose a protocol that includes spatially-distributed but infrequent sampling combined with a limited number of temporally intensive measurements.

11.7.2.1 Spatially Distributed Measurements

Within an individual plot, collars should be positioned using an unbiased method. We generally select a direction and distance from the plot center. Large logs and rocks should be avoided, as they make getting a seal difficult. If large logs and rocks are a substantial fraction of the surface (>10%), their area should be determined and the area represented by the collar samples adjusted during extrapolation because CO_2 will not diffuse through rocks or logs and will surface elsewhere. The number of collars that are necessary per subplot or plot will depend on the average variability within the subplots or plots. At our subalpine rocky mountain sites, we have observed within subplot coefficient of variability (CV) of approximately 26.7%. This suggests that, on average, seven collars would be required to get standard errors down to 10% (a somewhat arbitrary precision threshold chosen that can be varied as desired) of the mean within individual subplots (Fig.11.2). When we grouped all four subplots together our results indicate a CV of 34%, indicating that 11 collars per plot would achieve standard errors at 10% of the mean (Fig.11.2). We measure 12 collars per plot (3 per subplot) and as a result of the above variability results, we consider plots to be the experimental unit for soil respiration.

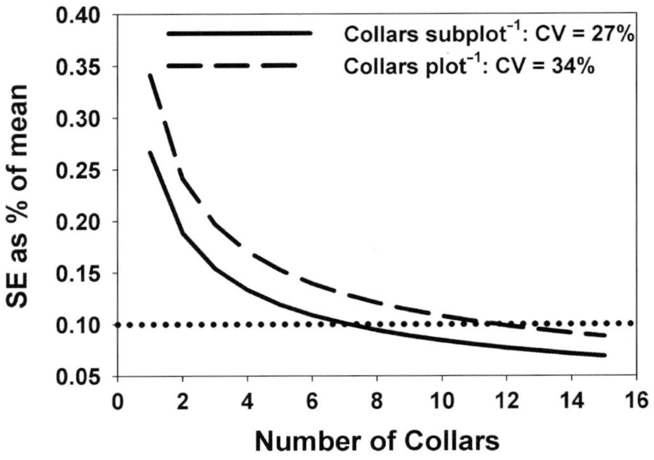

Fig. 11.2 Relationship between the number of collars within subplots or plots and soil respiration estimation accuracy, expressed as the standard error of individual collar measurements within either subplots (dotted line) or plots (solid line). Dotted line denotes SE = 10% of the mean

When attempting to estimate soil respiration for a large area, the purpose of plot locations is to characterize the spatial variation in soil respiration rates. These measurements should ideally span all the gradients in drivers of spatial variability. For example, we ensure that our plots include all the major plant communities and forest age categories and cover the elevation and aspect differences in the landscape. The number of plots required to characterize the entire landscape depends on the variability between plots. When treated independently, our subplots displayed a CV of 25%, suggesting that, if subplots were sufficiently sampled, only six subplots would be necessary to achieve a SE of 10% of the mean across the entire landscape (Fig. 11.3). We observed a CV of 18% between plots, indicating that three plots (or the equivalent of 12 subplots) would be required to achieve a SE of 10% of the mean (Fig. 11.3). At these spatially distributed locations, measurements should be collected at least several times throughout the year to get a reasonable measure of the relative respiration rate between points. We measure soil respiration once per month during the snow-free season and only 2–3 times during the winter when relatively consistent under-snow conditions create stable respiration rates.

11.7.2.2 Temporally Intensive Measurements

The purpose of these measurements is to characterize the temporal variability in soil respiration. They should be frequent enough to capture at least the diurnal fluctuations in soil respiration, and may need to be sampled at even higher frequency. In systems where pulse weather events are important, occasional diurnal

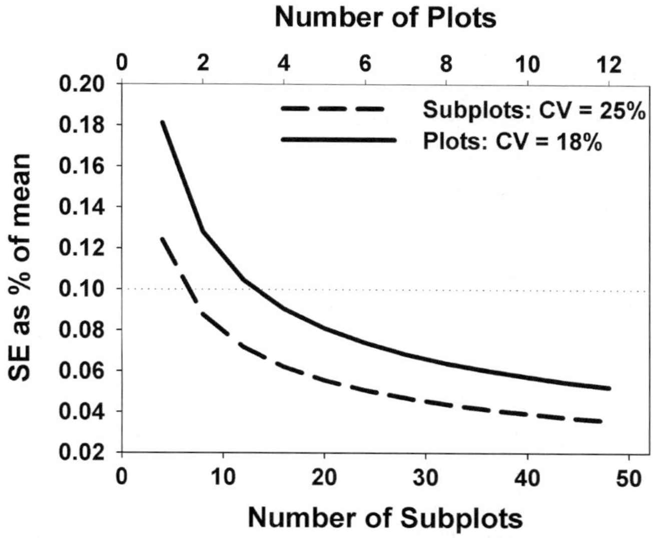

Fig. 11.3 Relationship between the number of plots or subplots and landscape-level soil respiration estimation accuracy, expressed as the standard error of subplots (dotted line) or plots (solid line). Dotted line denotes SE = 10% of mean

measurements are likely to miss the important consequences of these pulses. Even in ecosystems with relatively consistent weather conditions, the magnitude of diurnal variability in soil respiration may fluctuate throughout the year. Consequently, we suggest at least two diurnal collections per year (or an automated system), each consisting of soil respiration measurements every 2–4 hours for 24 hours. Placement of the diurnal collections within the landscape will depend on the sources of spatial variability, and should be located to span the major sources of spatial variability. For example, in a system where spatial variation is driven primarily by forest versus meadow, diurnal collections should sample from both forest and meadow. If elevation is the primary spatial driver, diurnal collections should be located at the top and bottom of the landscape. We utilize an automated system that samples from eight chambers divided between either forest and meadow or young and old stands, depending on the landscape.

11.8 Conclusions

Quantifying soil respiration at landscape scales is complicated by both the difficulties of accurately measuring soil respiration at a specific time and place and by the spatial and temporal variability inherent in soil respiration. Although multiple protocols for individual measurements have been developed, we recommend utilizing closed chambers with volume estimation by standard addition. This protocol is cost effective, highly portable and generates consistent results. Variability in soil respiration rates is attributable to environmental conditions, vegetation composition and abundance, and soil substrate quality. We maintain that accurate estimates of annual soil respiration at large scales will require a sampling strategy that captures both the spatial and temporal components of variation. We recommend a landscape-level sampling protocol that is a practical approach to capture important variability of soil respiration. The temporally continuous measurements at a few locations can be converted into annual soil respiration estimates, which, when combined with the infrequent, spatially distributed measurements can be scaled to the entire landscape.

Literature Cited

Adachi M, Bekku YS, Konuma A, Kadir WR, Okuda T, Koizumi H (2005) Required sample size for estimating soil respiration rates in large areas of two tropical forests and of two types of plantation in Malaysia. Forest Ecology and Management 210:455–459

Aikio S, Vare H, Strommer R (2000) Soil microbial activity and biomass in the primary succession of a dry heath forest. Soil Biology and Biochemistry 32:1091–1100

Amthor JS (2000) Direct effect of elevated CO_2 on nocturnal in situ leaf respiration in nine temperate deciduous tree species is small. Tree Physiology 20:139–144

Anderson JM (1973) Breakdown and decomposition of sweet chestnut (*Castanea sativa* Mill) and beech (*Fagus sylvatica* L) leaf litter in 2 deciduous woodland soils.1. Breakdown, leaching and decomposition. Oecologia 12:251–274

Bain WG, Hutyra L, Patterson DC, Bright AV, Daube BC, Munger JW, Wofsy SC (2005) Wind-induced error in the measurement of soil respiration using closed dynamic chambers. Agricultural and Forest Meteorology 131:225–232

Bolstad PV, Vose JM (2005) Forest and pasture carbon pools and soil respiration in the southern Appalachian mountains. Forest Science 51:372–383

Brooks PD, Schmidt SK, Williams MW (1997) Winter production of CO_2 and N_2O from alpine tundra: environmental controls and relationship to inter-system C and N fluxes. Oecologia 110:403–413

Brooks PD, Williams MW, Schmidt SK (1996) Microbial activity under alpine snowpacks, Niwot Ridge, Colorado. Biogeochemistry 32:93–113

Burton DL, Beauchamp EG (1994) Profile nitrous oxide and carbon dioxide concentrations in a soil subject to freezing. Soil Science Society of America Journal 58:115–122

Butnor JR, Johnsen KH (2004) Calibrating soil respiration measures with a dynamic flux apparatus using artificial soil media of varying porosity. European Journal of Soil Science 55:639–647

Butnor JR, Johnsen KH, Maier CA (2005) Soil properties differently influence estimates of soil CO_2 efflux from three chamber-based measurement systems. Biogeochemistry 73:283–301

Butnor JR, Johnsen KH, Sanchez FG (2006) Whole-tree and forest floor removal from a loblolly pine plantation have no effect on forest floor CO_2 efflux 10 years after harvest. Forest Ecology and Management 227:89–95

Campbell JL, Law BE (2005) Forest soil respiration across three climatically distinct chronosequences in Oregon. Biogeochemistry 73:109–125

Chambers JQ, Tribuzy ES, Toledo LC, Crispim BF, Higuchi N, dos Santos J, Araujo AC, Kruijt B, Nobre AD, Trumbore SE (2004) Respiration from a tropical forest ecosystem: partitioning of sources and low carbon use efficiency. Ecological Applications 14:S72–S88

Chen WJ, Chen J, Cihlar J (2000) An integrated terrestrial ecosystem carbon-budget model based on changes in disturbance, climate, and atmospheric chemistry. Ecological Modelling 135:55–79
Conen F, Smith KA (1998) A re-examination of closed flux chamber methods for the measurement of trace gas emissions from soils to the atmosphere. European Journal of Soil Science 49:701–707
Davidson EA, Savage K, Bolstad P, Clark DA, Curtis PS, Ellsworth DS, Hanson PJ, Law BE, Luo Y, Pregitzer KS, Randolph JC, Zak D (2002a) Belowground carbon allocation in forests estimated from litterfall and IRGA-based soil respiration measurements. Agricultural and Forest Meteorology 113:39–51
Davidson EA, Savage K, Verchot LV, Navarro R (2002b) Minimizing artifacts and biases in chamber-based measurements of soil respiration. Agricultural and Forest Meteorology 113:21–37
de Jong ED, Schapper HJ (1972) Calculation of soil respiration and activity from CO_2 profiles in soil. Soil Science 113:328–333
Dilustro JJ, Collins B, Duncan L, Crawford C (2005) Moisture and soil texture effects on soil CO_2 efflux components in southeastern mixed pine forests. Forest Ecology and Management 204:85–95
Edwards EW (1992) Likelihood. Johns Hopkins University Press, Baltimore, MD
Ekblad A, Bostrom B, Holm A, Comstedt D (2005) Forest soil respiration rate and delta C13 is regulated by recent aboveground weather conditions. Oecologia 143:136–142
Fahey TJ, Tierney GL, Fitzhugh RD, Wilson GF, Siccama TG (2005) Soil respiration and soil carbon balance in a northern hardwood forest ecosystem. Canadian Journal of Forest Research 35:244–253
Fang C, Moncrieff JB (1996) An improved dynamic chamber technique for measuring CO_2 efflux from the surface of soil. Functional Ecology 10:297–305
Gianfrani L, Rocco A, Battipaglia G, Castrillo A, Gagliardi G, Peressotti A, Cotrufo MF (2004) Assessing soil respiration by means of near-infrared diode laser spectroscopy. Applied Spectroscopy 58:1051–1056
Giardina CP, Ryan MG (2002) Total belowground carbon allocation in a fast-growing eucalyptus plantation estimated using a carbon balance approach. Ecosystems 5:487–499
Hanson PJ, Edwards NT, Garten CT, Andrews JA (2000) Separating root and soil microbial contributions to soil respiration: a review of methods and observations. Biogeochemistry 48:115–146
Hibbard KA, Law BE, Reichstein M, Sulzman J (2005) An analysis of soil respiration across northern hemisphere temperate ecosystems. Biogeochemistry 73:29–70
Hubbard RM, Ryan MG, Elder K, Rhoades CC (2005) Seasonal patterns in soil surface CO_2 flux under snow cover in 50 and 300 year old subalpine forests. Biogeochemistry 73:93–107
Janssens IA, Kowalski AS, Longdoz B, Ceulemans R (2000) Assessing forest soil CO_2 efflux: an in situ comparison of four techniques. Tree Physiology 20:23–32
Jassal R, Black A, Novak M, Morgenstern K, Nesic Z, Gaumont-Guay D (2005) Relationship between soil CO_2 concentrations and forest-floor CO_2 effluxes. Agricultural and Forest Meteorology 130:176–192
Kabwe LK, Farrell RE, Carey SK, Hendry MJ, Wilson GW (2005) Characterizing spatial and temporal variations in CO_2 fluxes from ground surface using three complimentary measurement techniques. Journal of Hydrology 311:80–90
King JA, Harrison R (2002) Measuring soil respiration in the field: An automated closed chamber system compared with portable IRGA and alkali absorption methods. Communications in Soil Science and Plant Analysis 33:403–423
King JS, Pregitzer KS, Zak DR, Sober J, Isebrands JG, Dickson RE, Hendrey GR, Karnosky, DF (2001) Fine-root biomass and fluxes of soil carbon in young stands of paper birch and trembling aspen as affected by elevated atmospheric CO_2 and tropospheric O_3. Oecologia 128:237–250
Kutsch WL, Liu CJ, Hormann G, Herbst M (2005) Spatial heterogeneity of ecosystem carbon fluxes in a broadleaved forest in northern Germany. Global Change Biology 11:70–88

Lavigne MB, Ryan MG, Anderson DE, Baldocchi DD, Crill PM, Fitzjarrald DR, Goulden ML, Gower ST, Massheder JM, McCaughey JH, Rayment M, Striegl RG (1997) Comparing nocturnal eddy covariance measurements to estimates of ecosystem respiration made by scaling chamber measurements at six coniferous boreal sites. Journal of Geophysical Research-Atmospheres 102:28977–28985

Law BE, Falge E, Gu L, Baldocchi DD, Bakwin P, Berbigier P, Davis K, Dolman AJ, Falk M, Fuentes JD, Goldstein A, Granier A, Grelle A, Hollinger D, Janssens IA, Jarvis P, Jensen NO, Katul G, Mahli Y, Matteucci G, Meyers T, Monson R, Munger W, Oechel W, Olson R, Pilegaard K, Paw KT, Thorgeirsson H, Valentini R, Verma S, Vesala T, Wilson K, Wofsy S (2002) Environmental controls over carbon dioxide and water vapor exchange of terrestrial vegetation. Agricultural and Forest Meteorology 113:97–120

Le Dantec V, Epron D, Dufrene E (1999) Soil CO_2 efflux in a beech forest: comparison of two closed dynamic systems. Plant and Soil 214:125–132

Liang NS, Inoue G, Fujinuma Y (2003) A multichannel automated chamber system for continuous measurement of forest soil CO_2 efflux. Tree Physiology 23:825–832

Liang NS, Nakadai T, Hirano T, Qu LY, Koike T, Fujinuma Y, Inoue G (2004) In situ comparison of four approaches to estimating soil CO_2 efflux in a northern larch (*Larix kaempferi* Sarg.) forest. Agricultural and Forest Meteorology 123:97–117

Litton CM, Ryan MG, Knight DH (2004) Effects of tree density and stand age on carbon allocation patterns in postfire lodgepole pine. Ecological Applications 14:460–475

Litton CM, Ryan MG, Knight DH, Stahl PD (2003) Soil-surface carbon dioxide efflux and microbial biomass in relation to tree density 13 years after a stand replacing fire in a lodgepole pine ecosystem. Global Change Biology 9:680–696

Lundegardh H (1927) Carbon dioxide evolution of soil and crop growth. Soil Science 23:417–453

Martin JG, Bolstad PV (2005) Annual soil respiration in broadleaf forests of northern Wisconsin: influence of moisture and site biological, chemical, and physical characteristics. Biogeochemistry 73:149–182

Massman WJ, Sommerfeld RA, Mosier AR, Zeller KF, Hehn TJ, Rochelle SG (1997) A model investigation of turbulence-driven pressure-pumping effects on the rate of diffusion of CO_2, N_2O, and CH_4 through layered snowpacks. Journal of Geophysical Research-Atmospheres 102:18851–18863

Massman WJ, Sommerfeld RA, Zeller K, Hudnell L, SG Rochelle (1995) CO_2 flux through a Wyoming seasonal snowpack: diffusional and pressure pumping effects. In: Biogeochemistry of snow-covered catchments. International Association of Hydrological Sciences, Wallingford. Pages 71–79

Murthy R, Griffin KL, Zarnoch SJ, Dougherty PM, Watson B, Van Haren J, Patterson RL, Mahato T (2003) Carbon dioxide efflux from a 550 m^3 soil across a range of soil temperatures. Forest Ecology and Management 178:311–327

Nay SM, Mattson KG, Bormann BT (1994) Biases of chamber methods for measuring soil CO_2 efflux demonstrated with a laboratory apparatus. Ecology 75:2460–2463

Norman JM, Garcia R, Verma SB (1992) Soil surface CO_2 fluxes and the carbon budget of a grassland. Journal of Geophysical Research-Atmospheres 97:18845–18853

Palmroth S, Maier CA, McCarthy HR, Oishi AC, Kim HS, Johnsen KH, Katul GG, Oren R (2005) Contrasting responses to drought of forest floor CO_2 efflux in a loblolly pine plantation and a nearby oak-hickory forest. Global Change Biology 11:421–434

Parkinson KJ (1981) An improved method for measuring soil respiration in the field. Journal of Applied Ecology 18:221–228

Pongracic S, Kirschbaum MUF, Raison RJ (1997) Comparison of soda lime and infrared gas analysis techniques for in situ measurement of forest soil respiration. Canadian Journal of Forest Research 27:1890–1895

Pumpanen J, Ilvesniemi H, Peramaki M, Hari P (2003) Seasonal patterns of soil CO_2 efflux and soil air CO_2 concentration in a Scots pine forest: comparison of two chamber techniques. Global Change Biology 9:371–382

Pypker TG, Fredeen AL (2002) Ecosystem CO_2 flux over two growing seasons for a sub-boreal clearcut 5 and 6 years after harvest. Agricultural and Forest Meteorology 114:15–30

Raich JW, Potter CS, Bhagawati D (2002) Interannual variability in global soil respiration, 1980–94. Global Change Biology 8:800–812

Rayment MB (2000) Closed chamber systems underestimate soil CO_2 efflux. European Journal of Soil Science 51:107–110

Rayment MB, Jarvis PG (2000) Temporal and spatial variation of soil CO_2 efflux in a Canadian boreal forest. Soil Biology and Biochemistry 32:35–45

Reichstein M, Rey A, Freibauer A, Tenhunen J, Valentini R, Banza J, Casals P, Cheng YF, Grunzweig JM, Irvine J, Joffre R, Law BE, Loustau D, Miglietta F, Oechel W, Ourcival JM, Pereira JS, Peressotti A, Ponti F, Qi Y, Rambal S, Rayment M, Romanya J, Rossi F, Tedeschi V, Tirone G, Xu M, Yakir D (2003) Modeling temporal and large-scale spatial variability of soil respiration from soil water availability, temperature and vegetation productivity indices. Global Biogeochemical Cycles 17:1104

Reichstein M, Subke JA, Angeli AC, Tenhunen JD (2005) Does the temperature sensitivity of decomposition of soil organic matter depend upon water content, soil horizon, or incubation time? Global Change Biology 11:1754–1767

Reiners WA (1968) Carbon dioxide evolution from the floor of 3 Minnesota forests. Ecology 49:471–483

Reth S, Reichstein M, Falge E (2005) The effect of soil water content, soil temperature, soil pH-value and the root mass on soil CO_2 efflux - a modified model. Plant and Soil 268:21–33

Rodeghiero M, Cescatti A (2005) Main determinants of forest soil respiration along an elevation/temperature gradient in the Italian Alps. Global Change Biology 11:1024–1041

Rustad LE, Huntington TG, Boone RD (2000) Controls on soil respiration: implications for climate change. Biogeochemistry 48:1–6

Ryan MG, Law BE (2005) Interpreting, measuring, and modeling soil respiration. Biogeochemistry 73:3–27

Schulze ED (1967) Soil respiration of tropical vegetation types. Ecology 48:652–653

Soegaard H, Nordstroem C, Friborg T, Hansen BU, Christensen TR, Bay C (2000) Trace gas exchange in a high-arctic valley 3. Integrating and scaling CO_2 fluxes from canopy to landscape using flux data, footprint modeling, and remote sensing. Global Biogeochemical Cycles 14:725–744

Sommerfeld RA, Mosier AR, Musselman RC (1993) CO_2, CH_4 and N_2O flux through a Wyoming snowpack and implications for global budgets. Nature 361:140–142

Sulzman EW, Brant JB, Bowden RD, Lajtha K (2005) Contribution of aboveground litter, belowground litter, and rhizosphere respiration to total soil CO_2 efflux in an old growth coniferous forest. Biogeochemistry 73:231–256

Takagi K, Nomura M, Ashiya D, Takahashi H, Sasa K, Fujinuma Y, Shibata H, Akibayashi Y, Koike T (2005) Dynamic carbon dioxide exchange through snowpack by wind-driven mass transfer in a conifer-broadleaf mixed forest in northernmost Japan. Global Biogeochemical Cycles 19

Tang JW, Baldocchi DD (2005) Spatial-temporal variation in soil respiration in an oak-grass savanna ecosystem in California and its partitioning into autotrophic and heterotrophic components. Biogeochemistry 73:183–207

Tang JW, Baldocchi DD, Qi Y, Xu LK (2003) Assessing soil CO_2 efflux using continuous measurements of CO_2 profiles in soils with small solid-state sensors. Agricultural and Forest Meteorology 118:207–220

Tang JW, Baldocchi DD, Xu L (2005a) Tree photosynthesis modulates soil respiration on a diurnal time scale. Global Change Biology 11:1298–1304

Tang JW, Misson L, Gershenson A, Cheng WX, Goldstein AH (2005b) Continuous measurements of soil respiration with and without roots in a ponderosa pine plantation in the Sierra Nevada Mountains. Agricultural and Forest Meteorology 132:212–227

Tate KR, Scott NA, Parshotam A, Brown L, Wilde RH, Giltrap DJ, Trustrum NA, Gomez B, Ross DJ (2000) A multi-scale analysis of a terrestrial carbon budget - is New Zealand a source or sink of carbon? Agriculture, Ecosystems and Environment 82:229–246

Underwood N, Hamback P, Inouye BD (2005) Large-scale questions and small-scale data: empirical and theoretical methods for scaling up in ecology. Oecologia 145:177–178

Wagner GH, Buyanovsky GA (1983) Use of gas sampling tubes for direct measurement of (CO_2)-C14 in soil air. International Journal of Applied Radiation and Isotopes 34:645–648

Wang W, Ohse K, Liu JJ, Mo WH, Oikawa T (2005) Contribution of root respiration to soil respiration in a C-3/C-4 mixed grassland. Journal of Biosciences 30:507–514

Wiant HV (1967a) Has contribution of litter decay to forest soil respiration been overestimated? Journal of Forestry 65:408–409

Wiant HV (1967b) Influence of moisture content on soil respiration. Journal of Forestry 65:902–903

Wiant HV (1967c) Influence of temperature on rate of soil respiration. Journal of Forestry 65:489–490

Widen B, Lindroth A (2003) A calibration system for soil carbon dioxide efflux measurement chambers: description and application. Soil Science Society of America Journal 67:327–334

Witkamp M (1966) Rates of carbon dioxide evolution from forest floor. Ecology 47:492–494

Xu LK, Baldocchi DD, Tang JW (2004) How soil moisture, rain pulses, and growth alter the response of ecosystem respiration to temperature. Global Biogeochemical Cycles 18:GB4002

Yim MH, Joo SJ, Shutou K, Nakane K (2003) Spatial variability of soil respiration in a larch plantation: estimation of the number of sampling points required. Forest Ecology and Management 175:585–588

Zheng DL, Chen JQ, LeMoine JM, Euskirchen ES (2005a) Influences of land-use change and edges on soil respiration in a managed forest landscape, WI, USA. Forest Ecology and Management 215:169–182

Zheng DL, Chen JQ, Noormets A, Euskirchen ES, Le Moine J (2005b) Effects of climate and land use on landscape soil respiration in northern Wisconsin, USA: 1972 to 2001. Climate Research 28:163–173

Chapter 12
Measurement of Methane Fluxes from Terrestrial Landscapes Using Static, Non-steady State Enclosures

Peter Weishampel and Randall Kolka

Abstract Wetlands are a dominant natural source of atmospheric methane (CH_4), a potent greenhouse gas whose concentration in the atmosphere has doubled over the past 150 years. Evaluating the impacts of CH_4 emissions on global climate and developing policies to mitigate those impacts requires a quantifiable and predictive understanding of natural CH_4 processing. Developing field sampling campaigns that quantify CH_4 flux in landscapes with prominent wetland features is a vital first step to developing that understanding. This chapter describes a field sampling approach that relies on static chambers to capture the CH_4 emitted from saturated soils and laboratory analyses of sequential samples to quantify CH_4 fluxes. Ultimately, by relating CH_4 fluxes from intensively sampled field sites to more easily measured ecosystem properties (e.g., temperature, water table, and productivity), models may be developed to predict CH_4 fluxes at larger landscape and regional scales.

Keywords Methane, saturated soils, static chamber, water table, wetland

12.1 Introduction

Atmospheric concentrations of methane (CH_4), the second-most abundant anthropogenic greenhouse gas, have doubled in the last 150 years (Wahlen 1993). Although atmospheric concentration of CH_4 is approximately 200 times less than that of carbon dioxide (CO_2), on a per-molecule basis, CH_4 is 25 times more effective

P. Weishampel
Department of Soil, Water and Climate, University of Minnesotam, 1991 Upper Buford Circle
St. Paul, MN 55108
E-mail: peter.weishampel@gmail.com

R. Kolka
US Forest Service, Northern Research Station, 1831 Highway 169 E. Grand Rapids,
MN 55744-3399
E-mail: rkolka@fs.fed.us

C.M. Hoover (ed.) *Field Measurements for Forest Carbon Monitoring*,

at trapping heat than CO_2. The rapid rise of atmospheric CH_4 coupled with its heat trapping properties highlight the importance of understanding CH_4 sources and sinks to help predict and manage the effects of CH_4 on the global climate system.

The net methane flux from terrestrial soils is comprised of competing microbial processes: methanogenesis (CH_4 production) and methanotrophy (CH_4 oxidation, or consumption). Methanogenesis is an anaerobic carbon mineralization pathway. Wetland soils, typically low in oxygen and often high in organic carbon, are the largest natural source for atmospheric methane (Schelsinger 1997). CH_4 produced in anoxic, saturated soils may be consumed in overlying unsaturated soils, affecting the net flux to the atmosphere. Additionally, dry upland soils can consume CH_4 that diffuses into soil from the ambient atmosphere (Born et al. 1990, Striegl et al. 1992). Given the importance of CH_4 in the carbon budget of some ecosystems and its strength as a greenhouse gas, it is considered an important component of the North American Carbon Program (Wofsy and Harriss 2002).

The processes influencing soil CH_4 fluxes are highly variable in space and time creating large uncertainty in the flux of CH_4 from terrestrial landscapes to the atmosphere. A hierarchical, scaled approach to improving our estimates of CH_4 fluxes from these landscapes is a high priority of the North American Carbon Program. This chapter describes a static-chamber approach to measuring CH_4 fluxes that is appropriate for spatially intensive sampling campaigns in landscapes with significant wetland components.

12.2 Measurement of Methane

Chamber-based approaches to trace gas measurement (e.g., such as those described by Bradford and Ryan in the previous chapter) offer the advantages of operational simplicity, portability, and low cost. An enclosure or chamber is placed over the soil to create a headspace of air, which can be sampled repeatedly over a short period of time (20–60 minutes). Samples are analyzed for target gas species, whose fluxes can be calculated based on the change in concentration over the duration of deployment. Unlike the infrared gas analyzers used to measure CO_2, portable instrumentation for providing instantaneous measurements of CH_4 in the field is not readily available, so gas samples must be stored and transported to a suitably equipped laboratory to measure CH_4 concentration.

12.2.1 Chamber Design and Deployment

Livingston and Hutchinson (1995) discuss factors to consider regarding the design, construction and deployment of chambers for measuring trace gas fluxes. For measuring CH_4 fluxes, the most commonly used technique utilizes a static vented enclosure system with two components: collars that are permanently installed in the ground and a portable chamber that fits over the collars. The use of permanently installed collars helps to create a tight seal with the soil surface while minimizing the disturbance effects of collar installation during periods of gas flux measurement. Chamber design and deployment described here is based upon approaches discussed in Livingston and Hutchinson (1995) and Holland et al. (1999). Here, chamber dimensions and design closely conform to those used for CO_2 measurements (Bradford and Ryan, Chapter 11) to take advantage of gas flux collars that were previously installed for that purpose. The primary difference between the chambers is that these static chambers contain no line hook-up to pumps, but rather, are designed for syringe sampling.

12.2.1.1 Materials

1. Permanently installed collars made of 10-in. diameter, schedule 40 PVC pipe, cut to 12.5 cm in length, with one edge beveled on a routing table.
2. Chamber components
   (a) Chamber top, made from 10-in. diameter, schedule 40, PVC cap. This should be light in color to minimize chamber heating during deployment (e.g., PlasticTrends part #D16010).
   (b) Vent, 1/4 in. stainless steel Swagelock fitting with 12 cm of plastic tubing (Livingston and Hutchinson (1995) discuss the optimum dimensions of vents as a function of chamber size and wind-speed).
   (c) Sample port, with either septum or valve, to accept syringe needle or luer-tip.
   (d) Thermocouple wire to monitor change in chamber air temperature during deployment period and thermocouple reader.
   (e) Inner PVC lip, made from 2 cm section of 10 -diameter PVC pipe, attached with PVC sealant to the inside of the cap, 2.5 cm from the base, with a self-adhesive closed cell foam gasket pressed to the lip bottom. During deployment, the gasket helps to create a seal against the PVC collar.
3. A plastic syringe (20 or 30 mL volume) with stopcock fitting and needle for sampling (25 gauge needles are commonly used). Extra syringes and needles are helpful to have with you in the field.
4. One 9-mL headspace vial with a septa and aluminum crimp-seal is needed for each sample taken. Exact vial dimensions may depend on the autosampler of the gas chromatagraph being used for the laboratory analysis. Vials should be sealed prior to visiting the field site. There are different strategies for preparing the vials. Vials can be sealed with laboratory air, provided that CH_4 concentration is low. Alternatively, vials can be flushed with nitrogen or evacuated with a vacuum pump. Because lab air may be variable in CH_4 content and evacuated vials may allow air to seep in prior to use, the nitrogen flushing approach may yield more consistently prepared vials. If the background gas in sealed vials is laboratory air, extra vials should be prepared as blanks to determine the

background concentration. Additionally, vials prepared in the lab to contain known concentrations of CH_4 should be taken to the field to evaluate potential leakage related to storage and transport.

An alternative approach to using vials for sample storage is to store the sample in the original syringe that is used to draw the sample from the chamber, sealing the sample with the stopcock. This approach eliminates possible errors in transferring gas samples to headspace vials, and eliminates time needed to prepare vials. If the available gas chromatograph makes use of an autosampler, this approach is not possible.

12.2.1.2 Collar Installation

1. Collars should be installed at least one week prior to sampling to minimize anomalies caused by soil disturbance.
2. Position collar on top of soil in desired location.
3. With long bread knife or saw or similar tool, cut around the edge of the collar
4. Press collar 5–10 cm into the soil until secure, leaving enough of the collar above the surface to enable the chamber to sit on the collar (~3 cm). In some cases, using a hammer to drive collars into the soil may expedite this operation, but this is not recommending for organic soils, where hammering can cause considerable compaction.
5. The depth from the collar top to the soil surface should be noted so that the actual volume of the enclosure can be calculated. Microtopographic variation in the soil may necessitate using the mean depth of multiple measurements.

Note: In organic or wet soils, the pressure on the soil from walking or standing near the collar can alter gas fluxes. Building boardwalks or platforms to stand on during sampling can help to minimize this disturbance.

12.2.1.3 Deployment

1. Measure and record the ambient air temperature.
2. Place the chamber on top of the collar to be measured so that the chamber gasket sits on the rim of the collar. Ground vegetation that interferes with the chamber seal should be tucked inside the chamber.
3. With a syringe, draw the initial sample (time = 0) and inject the sample into storage vials. The volume of sample injected should over-pressurize the vials to minimize leakage into the vial during sample storage. For 9-mL headspace vials, 15-mL injections of sample are typical. If using evacuated vials, experiment to evaluate the maximum sample volume that you can consistently inject to over pressurize the vial, and use that volume.
4. Sample the chamber again at 10, 20, and 30 minutes, injecting samples into new vials.

5. Depending on the proximity of other collars and the number of chambers available, simultaneous sampling of multiple chambers may be possible.
6. When the last sample is drawn, measure the chamber air temperature with the thermocouple reader and record.
7. Return samples to laboratory for analysis as soon as possible.

Figures 12.1 and 12.2 illustrate chamber design and deployment.

12.2.2 Laboratory Analysis

Vials should be analyzed for CH_4 using a gas chromatograph (GC) with a flame ionization detector (FID). There is a diversity of possible GC configurations for measuring CH_4 and they will not be covered here. Crill et al. (1995) and Holland et al. (1999) provide relevant information for configuring a GC to analyze CH_4.

Fig. 12.1 Gas sampling chamber positioned on collar with syringe equipped with stopcocks attached to 1/4″ bev-a-line tubing sample port. A 1/4″ diameter Swagelok fitting (on the right) serves as a vent

Fig. 12.2 Underside of gas sampling chamber adjacent to collar

12.2.3 Calculations

Flux calculations are based on the change in chamber CH_4 concentration over time. Concentrations, as measured by gas chromatography and adjusted as necessary for dilution in the headspace vial, can be converted into mass using the Ideal Gas Law:

$$m_{CH4} = (C_{CH4} \times M \times P \times V) / (R \times T) \qquad (12.1)$$

where

m_{CH4} = mass of CH_4-C in grams
C_{CH4} = concentration fraction of CH_4 (e.g., 2 ppm CH_4 as determined by chromatographic analysis = 2 x 10^{-6})
M is the molecular weight of C, 12.0107 g mol^{-1}
P is the pressure in atmospheres (~1)
V is the enclosure volume in L. Calculate as π x r^2 x D, where r is the chamber radius and D is the height of the chamber cap plus the mean depth from the collar rim to the soil surface.
T is the chamber air temperature when the gas was sampled.
R is the universal gas constant, 0.0820575 L atm K^{-1} mol^{-1}

For each collar, the mass of CH_4 should be plotted versus time to evaluate linearity of methane production. The rate of production (mass per time) is the slope of the best fit line through these points. The concentration may level off over time, causing a deviation from linearity. Such deviations are commonly attributed to artificial effects that stem from limitations of enclosures, and are typically discarded from the regression.

Calculate flux, the movement of mass through an area per unit time per unit time as:

$$f = a / A \qquad (12.2)$$

where

a = the slope of the best fit line described above and
A = the cross-sectional area of the collars.

12.3 Scaling CH_4 Fluxes

Measurements of CH_4 fluxes from wetland soils typically have high variability both spatially (i.e., collar to collar) and temporally (i.e., within collar). The number of chamber measurements needed to characterize seasonal or annual fluxes with confidence can be high (e.g., 50–100 individual collars per site; Holland et al. 1999). The labor required to carry out this degree of sampling is seldom available for individual sites, and is unrealistic for large landscapes or regions. Ultimately, seasonal and annual estimates of landscape- and regional-scale CH_4 fluxes may need to rely on modeling or remote sensing approaches that link methane fluxes to other ecosystem properties or processes that are more easily measured. Soil temperature, water table depth, and the spatial extent, community structure, and net primary productivity of wetland ecosystems have been used in regional methane flux models (e.g., Potter 1997, Potter et al. 2006). CH_4 flux datasets generated from intensive sampling campaigns and coupled with other data on ecosystem properties and processes should be highly valuable for developing and fine-tuning models of CH_4 flux at larger scales.

Literature Cited

Born M, Dorr H, Levin I (1990) Methane consumption in aerated soils of the temperate zone. Tellus 42B: 2–8

Crill PM, Butler JH, Coope, DJ, Novelli PC (1995) Standard analytical methods for measuring trace gases in the environment. In: Matson PA, Harriss RC (Eds.) Biogenic trace gases: measuring emissions from soil and water. Methods in Ecology Series, Blackwell Science, Oxford

Holland EA, Robertson GP, Greenberg J, Groffman PM, Boone RD, Gosz JR (1999) Soil CO_2, N_2O, and CH_4 exchange. In: Robertson GP, Coleman DC, Bledsoe CS, Sollins P (Eds.) Standard soil methods for long-term ecological research. Oxford University Press, New York

Livingston GP, Hutchinson GL (1995) Enclosure based measurement of trace gas exchange: applications and sources of error. In: Matson PA, Harriss RC (Eds.) Biogenic trace gases: measuring emissions from soil and water. Methods in Ecology Series, Blackwell Science, Oxford

Potter CS (1997) An ecosystem simulation model for methane production and emission from wetlands. Global Biogeochemical Cycles 11: 495–506

Potter C, Klooster S, Hiatt S, Fladeland M, Genovese P, Gross P (2006) Methane emissions from natural wetlands in the United States: satellite-derived estimation based on ecosystem carbon cycling. Earth Interactions 10: Article 22 1–12

Schelsinger WH (1997) Biogeochemistry: an analysis of global change. Academic Press, San Diego, CA

Striegl RG, McConnaughey TA, Thorstenson DC, Weeks EP, Woodward JC (1992) Consumption of atmospheric methane in desert soils. Nature 357: 145–47

Wahlen M (1993) The global methane cycle. Annual Review of Earth and Planetary Sciences 21: 407–426

Wofsy SC, Harriss RC (2002) The North American Carbon Program (NACP). Report of the NACP Committee of the U.S. Interagency Carbon Cycle Science Program. US Global Change Research Program, Washington, DC

Chapter 13
Measurement and Importance of Dissolved Organic Carbon

Randall Kolka, Peter Weishampel, and Mats Fröberg

Abstract The flux of dissolved organic carbon (DOC) from an ecosystem can be a significant component of carbon (C) budgets especially in watersheds containing wetlands. Although internal ecosystem cycling of DOC is generally greater than the fluxes to ground or surface waters, it is the transport out of the system that is a main research focus for carbon accounting. In watersheds containing organic wetland soils or peatlands, the flux from the watershed can be 4–8% of annual net primary production, a significant fraction that should be addressed when performing a carbon mass balance. Recent literature suggests that DOC transport from watersheds is increasing as a result of climate change or changes in sulfur deposition. As changes occur in land use, atmospheric deposition, and climate, response variables such as DOC will become even more critical to document the effect of those changes.

Keywords DOC, forest floor, soil carbon, TOC analyzer, wetlands

13.1 Introduction

Dissolved organic carbon (DOC) is operationally defined as organic molecules that pass through a filter, most often 0.45 μm. This is usually the major form of carbon transported with soil solution and in streams. The importance of DOC lies in its role

R. Kolka
US Forest Service, Northern Research Station, 1831 Highway 169 E., Grand Rapids, MN 55744-3399
E-mail: rkolka@fs.fed.us

P. Weishampel
Department of Soil, Water and Climate, University of Minnesota, 1991 Upper Buford Circle, St. Paul, MN 55108
E-mail: peter.weishampel@gmail.com

M. Fröberg
Department of Energy, Oak Ridge National Laboratory, Oak Ridge, TN, Current address: Department of Forest Soils, Swedish University of Agricultural Sciences, PO Box 7001, SE 750 07 Uppsala, Sweden
E-mail: mats.froberg@sml.slu.se

C.M. Hoover (ed.) *Field Measurements for Forest Carbon Monitoring*,

of being able to hydrologically transport carbon between different pools in the ecosystem. Most significant is the transport from the forest floor to the mineral soil. The internal fluxes of DOC within the ecosystem are in general higher than the net loss of DOC to ground water and surface waters.

DOC concentrations in rain water are generally very low but increase as the water passes through the canopy and forest floor. Fluxes of DOC in throughfall of temperate forests range from 4–16 g m^{-2} $year^{-1}$, whereas the flux in the O horizon is usually is in the range 10–40 g m^{-2} $year^{-1}$ (Michalzik et al. 2001). In the mineral soil DOC concentrations and fluxes decrease with depth and under the B horizon the flux is usually well below 10 g m^{-2} (Michalzik et al. 2001). The difference between O and B horizons is widely thought to be mainly due to physical and chemical retention rather than rapid mineralization (Kalbitz et al. 2000).

DOC transport in runoff increases with increasing proportion of wetlands present in the watershed, especially with organic soil wetlands or peatlands present (Aitkenhead et al. 1999). DOC exiting peatlands can be upwards of 4–8% of annual net primary productivity. Fluxes of DOC from watersheds containing wetlands typically range from 2–10 g m^{-2} $year^{-1}$ (Kolka et al. 1999, Elder et al. 2000). In watersheds or sites with few wetland soils, the loss of DOC is minimal in relation to other carbon pools and fluxes. In these systems, the error associated with measuring larger carbon fluxes is probably greater than DOC fluxes.

DOC fluxes are small compared to some other carbon fluxes in the ecosystem, but DOC may be important for carbon balances of litter and the O horizon. In relation to the annual aboveground litter fall, the annual transport of DOC from the O horizon to the mineral soil is on average 17%, with a range from 6–30% in temperate forests (Michalzik et al. 2001). DOC is also a significant source of organic carbon in the mineral soil (e.g. Neff and Asner 2001).

13.2 Sample Collection

Dissolved organic carbon is typically measured at either the plot or watershed scale. At the plot scale lysimeters, wells or piezometers are used. Lysimeters are typically used in unsaturated soils while wells and piezometers are used where water tables are present. There are two main types of lysimeters, zero tension and tension lysimeters. There are some differences between these two that need to be taken into consideration, as both the quantity and sources of DOC may be different, depending on the type of lysimeter used. Zero-tension lysimeters better reflect water that is moving through soils, as they mainly collect water in large pores. However, they create a discontinuity in the soil pore system and require the build up of a temporary local water table before they start collecting water. Therefore it may, in areas with moderate rainfall intensities, be difficult to get a sample of the soil solution with zero-tension lysimeters, at least in mineral soil. For minimum disturbance it is recommended that zero-tension lysimeters are installed laterally from pits, rather than by cutting through the forest floor from above.

Tension lysimeters consist of a porous cup connected to a collection flask where vacuum has been applied. Tension lysimeters are in contact with the soil pore system and collect soil water representing smaller pores than zero-tension lysimeters. They are more likely than zero-tension lysimeters to collect water in the mineral soil, but may be less representative of the DOC that is actually moving through the soil. Tension lysimeters may be installed laterally from pits or at an angle from the surface. Zero-tension lysimeters installed under the O horizon and tension lysimeters installed at depth in the mineral soil are often used in combination and this may be the best solution in many situations.

The depth at which the lysimeters should be installed depends on the question being asked. One set of lysimeters is often installed under the O layer to capture the flux of carbon at the interface between organic and mineral soil horizons. Another set of lysimeters is often installed in or under the B horizon and may represent the flux of DOC leaving the ecosystem.

In saturated conditions, wells or piezometers are commonly used to sample soil solution, including for the analysis of DOC. Wells are slotted their entire length and give a representative sample of the entire depth of the well (i.e. the soil profile). Piezometers are slotted only at the bottom and are used to sample a specific location, depth or horizon in the soil. Typically in studies, lysimeters, wells and piezometers are used in combination in upland to wetland transects to assess soil water concentrations and fluxes of DOC (McLaughlin et al. 1994).

Sampling of soil solution for DOC analysis may occur at fixed time intervals or based on precipitation events. Samples should be filtered after collection and kept refrigerated until analysis. For lysimeters, wells, and piezometers it is necessary to wait a couple of months to let the instruments equilibrate with their surroundings before samples are collected and analyzed.

At the watershed scale, samples are typically collected at the watershed outlet as grab samples or with automated equipment. Sampling is typically either event based or on a fixed interval. DOC concentrations vary with water flux in the stream and high fluxes of water are often combined with high concentrations of DOC and it is thus important to take samples during these events.

13.3 Measurement of DOC Concentration

Before DOC analysis the samples need filtration. The most commonly used pore size is 0.45 μm, but 0.2 μm and 0.7 μm are also common. Membrane filters are most commonly used, but syringe filters may be more convenient if small amounts of water are being filtered. Different kinds of filters are used, but cellulose acetate filters are probably most common. Most important is that the filters do not release any DOC during filtration. Samples from the mineral soil collected with tension lysimeters and samples from stream water may not always need filtering, but this needs to be evaluated for each site.

Numerous analyzers exist on the market, most of which are termed TOC analyzers. Measurement of DOC entails removing inorganic carbon with acid, sparging the resultant CO_2 and oxidizing the remaining C (presumably all OC) and measuring the CO_2 generated by the oxidation process. Oxidation of DOC can be accomplished by combustion, UV persulfate oxidation, ozone, or through UV fluorescence. Good reviews of the analytical methods can be found in Bolan et al. (1996) and Doyle et al. (2004).

13.4 Calculation of Fluxes

DOC fluxes are simply calculated by multiplying DOC concentration by the water flux; however, at the plot scale, probably the most difficult measurement is the flux of water. Due to disturbed hydrology, it is usually not possible to use the amounts of water collected in the lysimeters to estimate the flux. Computer models are sometimes used to estimate water fluxes. The measurement of soil moisture and hydraulic conductivity is one method to estimate the flux of water through the rooting zone. Micrometeorological techniques can also be applied by measuring surface inputs of precipitation or throughfall, estimating surface outputs in the form of evapotranspiration (ET) through the energy balance and measuring changes in soil moisture. Transport through the rooting zone can be calculated by difference using the hydrological mass balance.

At the watershed scale, typically flow is either measured by a device such as a weir or flume that is at the outlet of the stream exiting the watershed (Kolka et al. 1999). Where such devices are unavailable, stream gauging is commonly employed. Stream gauging entails measuring flow and relating flow to stream water height or stage height. Regression relationships are developed over a range of flows relating stage height to flow (Brooks et al. 2003).

Measurement of water fluxes to lakes and wetlands typically entails the use of groundwater wells and piezometers that measure the head of water upslope of the water body which allows for the estimation of inputs in saturated soil zones if one knows soil hydraulic conductivity (e.g. Freeze and Cherry 1979, Kolka et al. 2000, Mann and Wetzel 2000).

13.5 Summary

Comparison of DOC fluxes among ecosystems, treatments (e.g. harvest) or over time can elucidate changes in ecosystem processes. The few studies assessing harvesting on stream DOC vary in the response. Harvesting generally increases soil temperatures (more decomposition, higher DOC) but also reduces redox status (higher water tables, lower DOC) (Tate and Meyer 1983). Studies that see increases relate DOC increases simply to flow increases (e.g. Hobbie and Likens

1973). A number of studies have demonstrated that the amount of wetlands, especially peatlands, controls watershed level transport of DOC in streams (e.g. Gergel et al. 1999, Kolka et al. 1999). If there are wetlands present in the watershed, that factor appears to overwhelm any vegetation management factor controlling DOC transport. A number of watersheds have been experiencing increases in DOC transport as a result of increasing temperatures from climate change (Freeman et. al. 2001); however, other studies indicate that decreases in atmospheric deposition of sulfur may be the cause of the increases (Evans et al. 2006). As a result of changes in land use, management practices, climate and atmospheric inputs, DOC will continue to be an important response variable as we strive to understand carbon storage and fluxes.

Literature Cited

Aitkenhead JA, Hope D, Billett MF (1999) The relationship between dissolved organic carbon in stream water and soil organic matter pools at different spatial scales. Hydrologic Processes 13: 1289–1302

Bolan NS, Baskaran S, Thiagarajan S (1996) An evaluation of the methods of measurement of dissolved organic carbon in soils, manures, sludges, and stream water. Communications in Soil Science and Plant Analysis 27: 2723–2737

Brooks KN, Ffolliott PF, Gregersen HM, DeBano LF (2003) Hydrology and the management of watersheds. Third edition. Iowa State Press, Ames, Iowa, 590 pp

Doyle A, Weintraub MN, Schimel JP (2004) Persulfate digestion and simultaneous colorimetric analysis of carbon and nitrogen in soil extracts. Soil Science Society of America Journal 68: 669–676

Elder JF, Rybicki NB, Carter V, Weintraub V (2000) Sources and yields of dissolved organic carbon in northern Wisconsin stream catchments with differing amounts of peatland. Wetlands 20: 113–125

Evans CD, Chapman PJ, Clark JM, Monteith DT, Cresser MS (2006) Alternative explanations for rising dissolved organic carbon export from organic soils. Global Change Biology 12: 2044–2053

Freeman C, Evans CD, Monteith DT, Reynolds B, Fenner N (2001) Export of organic carbon from peat soils. Nature 412: 785–785

Freeze RA, Cherry JA (1979) Groundwater. Prentice-Hall, Englewood Cliffs, NJ

Gergel SE, Turner MG, Kratz TK (1999) Dissolved organic carbon as an indicator of the scale of watershed influence on lakes and rivers. Ecological Applications 9: 1377–1390

Hobbie JE, Likens GE (1973) The output of phosphorus, dissolved organic carbon and fine particulate carbon from Hubbard Brook watersheds. Limnology and Oceanography. 18(5): 734–742

Kalbitz K, Solinger S, Park J-H, Michalzik B, Matzner E (2000) Controls on the dynamics of dissolved organic matter in soils: a review. Soil Science 165: 277–304

Kolka RK, Grigal DF, Verry ES, Nater EA (1999) Mercury and organic carbon relationships in streams draining forested upland/peatland watersheds. Journal of Environmental Quality 28: 766–775

Kolka RK, Singer JH, Coppock CR, Casey WP, Trettin CC (2000) Influence of restoration and succession on bottomland hardwood hydrology. Ecological Engineering 15:131–140

Mann CJ, Wetzel RG (2000) Hydrology of an impounded lotic wetland – subsurface hydrology. Wetlands 20: 33–47

McLaughlin JW, Lewin JC, Reed DD, Trettin CC, Jurgensen MF, Gale MR (1994) Soil factors related to dissolved organic carbon concentrations in a black spruce swamp, Michigan. Soil Science 158: 454–464
Michalzik B, Kalbitz K, Park J-H, Solinger S, Matzner E (2001) Fluxes and concentrations of dissolved organic carbon and nitrogen – a synthesis for temperate forests. Biogeochemistry 52: 173–205
Neff JC, Asner GP (2001) Dissolved organic carbon in terrestrial ecosystems: synthesis and a model. Ecosystems 4: 29–48
Tate CM, Meyer JL (1983) The influence of hydrologic conditions and successional state on dissolved organic carbon export from forested watersheds. Ecology 64: 25–32

Part V
Supplemental Variables for Carbon Cycle Modeling

Chapter 14
Forest Canopy Structural Properties

Marie-Louise Smith, Jeanne Anderson, and Matthew Fladeland

Abstract The forest canopy is the interface between the land and the atmosphere, fixing atmospheric carbon into biomass and releasing oxygen and water. The arrangement of individual trees, differences in species morphology, the availability of light and soil nutrients, and many other factors determine canopy structure. Overviews of approaches for basic measurements of canopy structure are presented including height (maximum stem height), the fraction of ground overlain by foliage (coverage), and the leaf area per unit ground area (leaf area index, LAI). These measures are important for characterizing estimates of ecosystem productivity and development, as well as for interpreting and validating landscape- to regional-scale estimates of vegetation attributes derived from remote sensing data. The approaches outlined are not exhaustive, but rather identify a range of robust ground- and remote sensing-based methods that capture a suite of metrics useful to the goals of the North American Carbon program: to provide robust, ground-based estimates of C uptake, storage, and flux that can be used for validation and scaling.

Keywords Direct measurement, forest structure, indirect measurement, remote sensing, scaling, leaf area index

---

M.-L. Smith
US Forest Service, Northern Research Station, Durham, NH, Current address: US Forest Service, Legislative Affairs, 201 14th Street, SW, Washington, DC 20250-1130
E-mail: marielouisesmith@fs.fed.us

J. Anderson
Complex Systems Research Center, University of New Hampshire, 39 College Road, Durham, NH 03824
E-mail: jeanne.anderson@unh.edu

M. Fladeland
NASA Ames Research Center, Mail Stop 242-2, Moffett Field, CA 94035
E-mail: matthew.fladeland@nasa.gov

C.M. Hoover (ed.) *Field Measurements for Forest Carbon Monitoring*,

14.1 Overview: Supplemental Variables for Carbon Cycle Modeling

The ability to efficiently extend site specific measurements to broader scales enhances our ability to examine key relationships associated with these traits at various levels from the site to the forest stand and, with remote sensing technologies, to larger landscapes. A growing body of research demonstrates the complementary nature (and necessity for integration) of distributed site measurements, remote sensing and ecosystem modeling in studies of terrestrial carbon cycling. Even very comprehensive field measurement campaigns cover only small portions of any landscape. Understanding landscape-scale and larger patterns is difficult, if not impossible, without some means of extending point-based field measurements. There is widespread acknowledgment that remote sensing holds a central role. Remote sensing instruments capture spatially continuous information on the reflectance properties of landscape and vegetation and are now capable of providing regular, high-resolution (10-m to 1-km) mapping and monitoring of land surface characteristics relevant to understanding and modeling of ecosystem processes (Turner et al. 2004). Variables that are now routinely retrieved include vegetation type and cover, biomass, stand age class, phenological status, leaf area index, tree height, and biochemistry (Ollinger et al. 2007). Integration of field-based and remotely-sensed data sets with ecosystem process models provides a means for landscape- to regional-scale assessment of carbon fluxes and analysis of the underlying processes affecting them. In this and the following chapter we present a summary of field methods useful in characterizing the structure and biochemistry of forest canopies, that are commonly used in terrestrial productivity estimation and modeling, and which may be effectively observed and scaled with remote sensing.

14.2 Introduction

The forest canopy is the interface between the land and the atmosphere, fixing atmospheric carbon into biomass and releasing oxygen and water. An understanding of canopy structure is important for modeling the transport and sequestration of atmospheric gases such as carbon dioxide, methane and even metals. The arrangement of individual trees, differences in species morphology, the availability of light and soil nutrients, and many other factors determine canopy structure, with differences contributing to variations in the amount of potential biomass and the availability of resources to understory communities.

Among the most common measurements of canopy structure are height (maximum stem height), the fraction of ground overlain by foliage (coverage), and the leaf area per unit ground area (leaf area index, LAI). Aspects of structure may also be used to predict stand attributes, such as stem density, basal area and aboveground biomass. These measures are important for characterizing estimates of ecosystem productivity and development, as well as for interpreting and validating landscape- to regional-scale estimates of vegetation attributes derived from remote sensing data.

The canopy, as Geoffrey Parker notes in his 1995 work on the structure and microclimate of forest canopies, is the aggregate of all tree crowns, inclusive of foliage and woody materials.

Canopy structure, in his words, is "the organization in space and time, including the position, extent, quantity, type and connectivity, of the aboveground components of vegetation." Many terms are used interchangeably in reference to canopy structure including: physiognomy (the shapes of individual crowns), architecture (the growth patterns and resultant forms of stems), organization (the statistical distribution of canopy components or important characteristics in space or time) and canopy texture (the size of crown units composing the overstory that are apparent from above the stand) (Parker 1995, Parker et al. 2004).

Different developmental stages of forest canopies yield various structures that use light resources in different ways. While recognizing the continuous nature of forest structural development and the arbitrary nature of classification schemes, Franklin et al. (2002) describe a series of stages in forest stand development. Stages of disturbance and legacy creation, cohort establishment, canopy closure, biomass accumulation/competitive exclusion, maturation, vertical diversification, horizontal diversification, and pioneer cohort loss are broadly applicable across varying biomes. Parker and Brown (2000) advocate for basic efforts to illustrate the distribution of relevant structures across space, maps of structure, function and microclimatic environment that illustrate ecological gradients within the canopy systems.

The sections that follow provide an overview of measurements of height, canopy cover and LAI of use to North American Carbon Program research. The main goal of the NACP sites is to provide robust, ground-based estimates of C uptake, storage, and flux that can be used for validation and scaling. The approaches outlined here are not exhaustive, but rather identify a range of robust ground- and remote sensing-based methods that capture a suite of useful metrics. It is critical to note, however, that such summary measures describe mean conditions; often providing little or no information on spatial variability. Parker et al. (2004) caution that "such summaries are often inadequate for prediction of growth, carbon dioxide exchange, structural complexity and habitat quality; structural information on the spatial distribution of canopy components is more useful than summary measures (Parker 1995, Lefsky et al. 2002)."

14.3 Canopy Height

Vegetation height is a function of species composition, climate, disturbance and site quality. Understanding the vertical distribution of vegetation is an important element in reducing the uncertainty in estimates of the net primary productivity of different ecosystems. When coupled with knowledge of composition, disturbance history, and other environmental factors, height often serves as an estimate of stand age or successional state (Dubayah et al. 2000). Height can be measured from the ground or by remote sensing (inclusive of aerial photography).

14.3.1 Remote Sensing Approaches

Remote sensing approaches to the measurement of height include interpretation of data from passive, optical sensors (e.g. Landsat TM, aerial photography) as well as active, physical sensors (e.g. radar, lidar). In recent years, lidar remote sensing has generated much attention as a breakthrough technology for deriving the three-dimensional characteristics of canopies, most notably height (Dubayah et al. 2000, Lefsky et al. 2002). Dubayah and his colleagues note, "that unlike visible, near-infrared, and radar remote sensing techniques that require sophisticated models to recover basic canopy structure, height is directly measured by lidar systems and, as long as the ground can be determined, is readily obtained." The direct measurements of waveform lidar are canopy height, sub-canopy topography and the vertical distribution of intercepted surfaces between the canopy top and the ground. Other biophysical attributes of forest structure, such as aboveground biomass, are modeled or inferred from these direct measurements, or by integration with data from other imaging sensors (optical or radar) (Dubayah et al. 2000, Lefsky et al. 2002, Treuhaft et al. 2002). Recent work with airborne multi-angle data has also provided results for remote measurements of height that correlate well with waveform lidar measures (Kimes et al. 2006).

Lidar data require some validation with ground measured field data and may require a modest calibration to optimize consistency with more traditional forestry estimates of average stand height. Imaging waveform lidar systems, such as NASA's Laser Vegetation Imaging Sensor (LVIS) (Blair et al. 1999), map continuous fields of heights providing important data on spatial variation of height across the landscape (Dubayah et al. 2000). Several approaches to validating remotely derived height from ground measures are discussed below.

14.3.2 Ground-Based Approaches

Ground-based laser scanning is an emerging technology with great potential for forest measurements and modeling on a smaller scale. Differing applications of ground-based laser scanning have provided high-resolution, spatially explicit measures of plot-level forest canopy structure (Henning and Radtke 2006, Parker et al. 2004, Lovell et al. 2003). Such measurements can provide improved assessments of canopy structure, better linking leaf-level processes to canopy structure and forest growth (Henning and Radtke 2006). These techniques can quantify spatial variation and are an advance over some previous methods, allowing application to long-term studies of vegetation change (Parker et al. 2004).

Laser rangefinders can obtain precise field height data and create detailed stand maps. A combination of direct (i.e. slope distance and inclination) and calculated (i.e. horizontal distance, vertical distance, and percent slope) measurements are used (e.g. Adams and Zuchowski 1997). Terrain and density can influence error and ease of use. Sonic hypsometers can also be used to collect height or horizontal distances. These instruments emit omni-directional sonic pulses that permit distance estimation (i.e. horizontal distance, vertical distance, and percent slope).

Clinometers are surveying instruments for measuring angles of elevation, slope, or incline. By using the principles of trigonometry, the height of tall objects is calculated from the angles measured. Measured poles are used to assess trees, generally of shorter stature, with excellent accuracy. These traditional methods are, however, typically quite time intensive.

Throughout its century-long history, forestry has relied on various types of tree volume and dry weight (aboveground biomass) tables (Avery and Burkhart 1983) to explain relationships between tree volume and diameter at breast height (DBH), tree height, and at times tree form. Height in these tables is typically expressed as total tree height or length of merchantable stem. Although many independent variables have been incorporated into regression equations for predicting tree volume, measurements of stem diameter and height tend to account for the greatest proportion of the variability in volume. Numerous height-diameter regression models have been proposed and used in the past and continue to evolve today. Because the basic height-diameter relationship is not linear, but is sigmoid in shape over the full range of diameters, transformation is required to apply linear regression methods for estimating coefficients in the relationship. Clark and Clark (2000) correctly caution, however, that "while tree height can be well predicted from DBH measurements, generalized allometric equations for a given life zone may not correctly describe the allometry of tree species at a particular study site."

14.4 Canopy Cover

Measurements of canopy cover can provide information on forest productivity, solar energy balance, the history of natural and anthropogenic disturbances, below canopy processes, stand density, and overall system change. Cover metrics are widely used as a means of defining forests and classifying forest types (e.g. the United Nations Food and Agriculture Organization (FAO 2000)) and are often considered in land and watershed management decision-making. Canopy cover has been defined by Korhonen et al. (2006) as the "proportion of the forest floor covered by the vertical projection of the tree crowns, and should be distinguished from canopy closure, which is defined as the proportion of sky hemisphere obscured by vegetation when viewed from a single point (Jennings et al. 1999)." In other words, "if canopy is measured with instruments that have an angle of view (i.e. measure a larger area than just a vertical point), like cameras (Kuusipalo 1985) or spherical densiometers (Cook et al. 1995), the results are estimates of canopy closure" (Korhonen et al. 2006).

As such, canopy cover measurements, by definition, should be made in the exact vertical direction (Jennings et al. 1999). If instruments with an angle of view are used, canopy cover is usually overestimated (Bunnell and Vales 1990, Cook et al. 1995, Jennings et al. 1999, Korhonen et al. 2006). In practice, however, if the angle of view is narrow (i.e. less than 30°) the bias is considered insignificant (Bunnell and Vales 1990). Another issue worth noting is that while tree height and length of the live crown do not affect the estimates of canopy cover, canopy closure increases as the trees become taller, and as the height to the live base of the crown decreases (Jennings et al. 1999). Canopy cover can be measured from the ground or with remote sensing (inclusive of aerial photography). It can also be modeled statistically from stand parameters (e.g. Forest Vegetation Simulator (FVS), Donnelly and Johnson 1997).

14.4.1 Remote Sensing Approaches

By virtue of their high spatial resolution, aerial photographs and digital imagery are often used for measuring canopy cover across a landscape, providing metrics at a resolution of 1–30 m, while generally lower resolution satellite imagery is used to measure landscape cover and change at the regional to global scale with pixel size ranging from 250 m–2 km.

The January 2007 release of the National Land Cover Dataset by the interagency Multi-Resolution Land Characteristics Consortium (MRLC) includes Landsat 7 (ETM+) - derived percent forest canopy density coverage at 30-m resolution for the conterminous United States. These products are web enabled for download from the MRLC website at http://www.mrlc.gov. Other NLCD 2001 products include 21 classes of land cover and percent urban imperviousness. The method employed to map tree canopy density for NLCD 2001 is based on empirical relationships between tree canopy density and Landsat data established using regression tree techniques. One-meter digital orthophoto quadrangles were used to derive tree canopy density data needed for calibrating the relationships between canopy density and Landsat spectral data. This method is described in detail in Huang et al. (2001).

Tree cover and canopy density estimates can be obtained from the MODIS (15 arc-seconds) Vegetation Continuous Fields (MODIS VCF) - Tree Cover Dataset. The Vegetation Continuous Fields collection contains proportional estimates for vegetative cover types: woody vegetation, herbaceous vegetation, and bare ground. The product is derived from all seven bands of the Moderate-Resolution Imaging Spectroradiometer (MODIS) sensor onboard NASA's Terra satellite. The continuous classification scheme of the VCF product may depict areas of heterogeneous land cover better than traditional discrete classification schemes. While traditional classification schemes indicate where land cover types are concentrated, this VCF product shows how much of a land cover such as "forest" or "grassland" exists anywhere on a land surface (Hansen et al. 2003). The tree cover estimates in this dataset are calculated using an automated algorithm for images acquired from the MODIS sensor over a one year period (October 2000– December 2001). This dataset is available online from the University of Maryland's Global Land Cover Facility (GLCF) at http://glcf.umiacs.umd.edu/data/modis/vcf/.

With waveform lidar, the relative strength of the canopy and ground returns provides information on canopy closure (Harding et al. 2001). In the case of LVIS, the relevant variable is the fraction of laser illumination that is returned to the sensor from the ground—the ground return (Lefsky 1997). Initial processing of the cover metric involves the calculation of a simple ground-return ratio of canopy energy to total energy in the lidar waveform (see Eq. 14.1). This ground-return ratio provides an approximation of the degree of canopy closure (Drake 2001).

$$CC = C_e / (C_e + G_e) \tag{14.1}$$

Where CC = canopy closure; C_e = total energy of canopy (in photon counts); G_e = total energy of ground (in photon counts) (Hofton et al. 2000).

14.4.2 Ground-Based Approaches

Visual estimations of canopy cover are point samples that typically determine the overhead presence or absence of canopy cover in the observed site. Sighting tubes (e.g. Cajanus tube) can help to ensure the verticality of the observation and improve the accuracy of results with suitable sample sizes (i.e. >100 points; Korhonen et al. 2006). Line-intercept sampling extends the concept by recording horizontal distances covered by live crown along a line-transect. Accuracies using these methods are generally high. While typically more time intensive than other methodologies, the line-intercept method is appropriate if more detailed information on layering and species is required (Fiala et al. 2006).

Various inexpensive hand-held devices (e.g. Moosehorn densiometers) can add additional rigor to visual estimates by calculating percentage cover at each sampled location. The Moosehorn, for example, uses an upward looking mirror with a grid that allows for the measurement of area for a given canopy. The recorder notes the number of open sky versus vegetation observations for a point field visible within the instrument viewer. Relative cover for the plot is the mean of the number of readings. This technique has been used to validate lidar metrics at the level of individual footprints (Hyde et al. 2005). Fiala et al. (2006) recommend the use of Moosehorn sampling for obtaining rapid, efficient estimates of vertically projected canopy cover.

Spherical densiometers are most frequently used for estimating canopy closure, but these estimates may have low accuracy and low precision (Strickler 1959). Although such problems can be reduced by use of a tripod, this does not eliminate the subjectivity involved in using these instruments. With their intermediate angle of view, densiometers are perhaps most accurately viewed as a hybrid estimate of cover and light, making them most useful where wide angle cover is the desired attribute to quantify (Fiala et al. 2006).

Another point sample method for measuring canopy cover and canopy closure is the upward looking imaging of a canopy using a camera with a hemispherical lens. The resultant image is a circle in which the center of the image is the zenith and the distance from center relates to view angles. When oriented and interpreted properly, these images can provide robust estimates of canopy cover, canopy closure, and canopy light interception as well as daily and yearly estimates of understory light, when overlaid with solar charts. Repeated measurements over time yield very detailed measurements of canopy closure and crown competition. When used with photon sensors this method can provide very accurate estimates of canopy light interception and when used with a robust sampling strategy, a useful estimate of total canopy cover through the gap fraction index and LAI (see below). Fiala et al. (2006) caution, however, that much hemispherical photography is really an estimate of light penetration, and is more appropriately used as a measure of angular canopy openness, rather than canopy cover.

Korhonen et al. (2006) and Fiala et al. (2006) provide recent comprehensive reviews and comparisons of ground-based approaches to estimation of canopy cover and closure and associated literature.

14.5 Leaf Area Index

Leaf area index (LAI) is one of the most challenging forest canopy structural metrics to quantify accurately, owing to large spatial and temporal variability and to measurement limitations inherent in all current methodologies. Numerous comprehensive reviews of theory, methodology, and instrumentation exist in the literature (e.g. Norman and Campbell 1989, Welles 1990, Fassnacht et al. 1994, Welles and Cohen 1996, Chen et al. 1997, Gower et al. 1999, Breda 2003, Jonckheere et al. 2004). A synthesis by Asner et al. (2003) provides some insight into the range in LAI observations globally across biomes, and the effect of more precise definition and improved measurement methodology on LAI estimation.

LAI is broadly defined as the amount of functional (green) leaf area in a canopy per unit ground area. Asner et al. (2003) describe four common measures of LAI that appear in the literature (Table 14.1). Of these, the most widely used is one-half the total green leaf area per unit ground surface area (definition 2, Chen and Black

Table 14.1 Leaf area index definitions (as reported in Asner et al. 2003)

1	Total	Total outside area of the leaves, taking leaf shape into account, per unit of horizontal ground area below the canopy
2	One-sided or one-half	One-half the total LAI (hemi-surface area or HSA), even if the two sides of the leaves are not symmetrical.
3	Horizontally projected	The area of 'shadow' that would be cast by each leaf in the canopy with a light source at infinite distance and perpendicular to the leaf surface, summed for all leaves in the canopy.
4	Inclined, projected	The projected area of leaves while accounting for leaf inclination.

1992) or the hemi-surface area (HSA, Gower et al. 1999). This definition is the most general measure for all leaf shapes, and is the appropriate parameter for radiation transfer models upon which indirect LAI measurement methods rely (Chen and Black 1992, Gower et al. 1999). This definition is most widely used in current literature and is the definition used in this chapter's discussion of approaches for estimating LAI.

The major methodologies for estimating LAI utilize either 'direct' approaches, such as destructive sampling, allometry, litterfall collection, and point contact sampling, or 'indirect' approaches involving optical instruments combined with modeling.

14.5.1 Direct Measurement

These approaches, particularly litterfall collection and destructive harvest, are commonly used as reference for evaluation or calibration of the more rapid, indirect methods and, in all cases, are best suited for estimation of maximum LAI. All direct methods involve determination of leaf area using either a leaf area meter (projected leaf area) or a specific relationship of dimension to area. For conifer species, projected leaf area must be adjusted by a coefficient related to the needle cross-sectional area (Chen and Black 1992, Chen et al. 1997).

In all cases green foliage area is measured on a sub-sample of leaves and related to dry foliage mass using dry weight to species-specific leaf area ratios, leaf mass per unit area (LMA, g/cm^2). The total dry mass of leaves collected within a known ground surface area is converted to LAI by multiplying the dry weight of each species by its species-specific LMA. The most critical step is accurate or representative determination of species-specific LMA, which varies vertically with the light environment in vegetative canopies, as well as in both space and time in response to resource availability. Rather than assessing LMA only from litterfall samples, LMA should be assessed on a site-specific basis from green leaf samples collected from several heights in the canopy (see "Foliar Chemistry" section for a method for LMA determination).

Harvest methods are the most accurate for direct estimation of LAI and involve destructive sampling and measurement of the area of all leaves within a delimited area. Harvest methods are most widely used in agricultural systems or those dominated by annual or small stature vegetation. In perennial, large stature systems such as forests, this method is challenging to implement and, in any case, precludes long-term study of LAI dynamics over time in a particular location. Most often destructive harvest methods in forests are used to develop site-specific allometric equations.

Allometric methods are based on physical dimensions, such as stem diameter at breast height (DBH) or sapwood area, using species-specific or stand-specific relationships derived from detailed destructive measurements of sub-samples of leaves, branches, and stems. As LAI is both spatially and temporally dynamic, allometric estimators of LAI are quite site specific and in any case cannot be used to capture temporal dynamics. Gower et al. (1999) advocate the development of site-specific allometric equations when accurate estimates of LAI are needed, particularly as reference for indirect methods used at the same site in the same period. Caution is advised when using existing allometric equations if site-specific equations cannot be developed, with particular attention paid to matching plant species and size when selecting such equations (Gower et al. 1999).

Point contact methods, which determine LAI from the mean contact number of a thin probe or plumb line that passes through the canopy (Warren-Wilson 1959), are accurate but are generally impractical in forest stands because of the tall stature of individuals and the often high density of leaves. A number of indirect optical methods have been developed based on this concept.

Litterfall collection is perhaps the most widely used non-destructive, direct method employed in forest ecology studies. Litterfall is typically collected in traps of known area distributed below the canopy before seasonal leaf fall. Collected litter is oven-dried and weighed to determine the litter dry weight (g m^{-2}). Leaf dry mass is converted to leaf area by multiplying the collected biomass, usually per species, by the species-specific LMA. Leaf area index is the accumulated leaf area over the period of leaf fall, which is sometimes seasonal but, more typically, the accumulated litterfall over one year. In deciduous systems, yearly litterfall (including small twigs, flowers, fruit, and seeds) is often used as an equivalent to yearly canopy production, minus herbivory. For a recommended approach, see Litterfall Methods in Aboveground Fluxes section (Bernier et al., this volume).

With adequate spatial and temporal sampling, determining leaf area index by litterfall collection is most appropriate in deciduous stands, which shed a majority of foliage during a single season. In needle-leaved evergreen and mixed forests, estimation of LAI through litterfall collection is more problematic as needle fall is not as strongly seasonal as that of deciduous species nor related to new growth, but rather related to leaf lifespan and the cumulative climate conditions over the leaf life span. In these cases, needle-leaved litterfall is often multiplied by average leaf retention time to approximate needle-leaved mass in the canopy.

14.5.2 Indirect Measurement

As direct measurement of LAI in forest ecosystems presents many practical and logistical difficulties, much effort has gone into devising more rapid, indirect methods. Optical, indirect measures afford the opportunity to monitor seasonal dynamics of LAI that is either not possible or practical with most direct measurement approaches.

The most widely used methods are based on radiative transfer theory and rely on the dependency between canopy element distribution (i.e. angle and spatial distribution) and canopy gap fraction and size distribution to assess forest canopy parameters, especially LAI. All optical instruments used to estimate LAI measure light transmittance. Measures obtained with optical instruments are termed 'effective LAI' or L_e, because these measures, based on Beer's law, are calculated assuming a random spatial distribution of foliage, and represent solar radiation interception without the distinction between woody and photosynthetically active materials.

The assumption of random spatial distribution of foliage is often not met in forest canopies, and strategies (and instrumentation) to account for non-random foliage distribution, as well as the contribution of non-photosynthetic material (stem, branches), must be employed to convert L_e to LAI (see Kucharik et al. 1999, Gower et al. 1999, Chen et al. 1997). Non-randomness in foliage distribution, or foliage clumping, can occur at leaf, branch, and individual tree levels (Gower et al. 1999). In forests, L_e is generally only 30–70% of the true LAI because of foliage clumping (Chen and Cihlar 1995). The clumping index, which is derived from canopy gap size estimates, generally pertains to clumping at scales larger than the defined element size in the canopy. For needle-leaved species, this is the shoot, for broad-leaved species this is the individual leaf. Non-random distribution of needles on conifer shoots cannot be measured with remote optical instruments, but must be measured by direct methods if "within-shoot" clumping is to be considered in optical measurement "clumping" corrections (see Gower et al. 1999).

If no correction for woody material is employed the optically-based measures are most properly plant area index (PAI) or surface area index (SAI) rather than LAI. In many forests, proportion of wood area to the total plant area is small; however, this proportion can be up to 30% in cases where LAI is very low such as in the boreal forest (Chen et al. 1997). One notion is to use estimates of wood area to correct plant area estimates. For example, in deciduous forests, measurements could be obtained at a given site before and after leaf senescence and the wood area index (WAI) could be subtracted from the PAI to yield LAI. However, some evidence suggests leaves preferentially mask branches (Kucharik et al. 1998) so simply subtracting wood area from plant area is likely overcompensation. Further, several studies have indicated that branches intercept only a small amount of light in healthy forest canopies (e.g. Fassnacht et al. 1994, Kucharik et al. 1998). However, additional correction may be needed for some forests such as those with partially defoliated canopies or canopies that retain a large amount of dead branches.

Indirect estimates of LAI plateau between 5–6, caused by gap fraction saturation as LAI increases (Gower et al. 1999). Direct measurement is the only reliable approach for canopies with LAI > 6. However, the global mean LAI as reported by Asner et al. (2003) is 4.5 (SD = 2.5) with maximum values rarely exceeding 7.0, except in forest plantations.

Two broad categories of optical instruments are commonly used to measure LAI: those that sample the canopy hemispherically, with LAI estimation based upon canopy gap fraction analysis, and those that sample along transects, with LAI estimation based upon canopy gap size distribution analysis.

14.5.2.1 Canopy Hemispherical Sampling

This approach quantifies canopy gap fraction at multiple zenith and azimuth angles obtained simultaneously, and assumes random foliage distribution. Measurements with these instruments are most reliable when performed under diffuse sky conditions (such as occur at dawn and dusk or under uniformly cloudy conditions) to minimize direct scattering of sunlight off leaves and trunks. In all cases correction factors, i.e. "clumping" factors, to account for non-random foliage distribution are also required to accurately convert L_e to PAI. However, if the goal of measurement is simply to characterize the light environment, or to assess relative differences in L_e over time and between stands, correction factors may not be needed.

Hemispherical (fish-eye) imaging is a technique to characterize plant canopies using images taken looking upward through an extreme wide-angle lens. Typically, an 180° 'fish-eye' camera lens coupled with a digital or film camera system is used to produce images of the projection of a hemisphere on a plane. The resulting circular image shows a complete view of all sky directions with the zenith in the center of the image and the horizons at the edges. These circular images record the size, shape, and location of gaps in the vegetation canopy from which such canopy structural parameters as openness, foliage inclination angle, and LAI may be calculated. Calculations of canopy metrics, such as leaf area index or canopy cover, depend on accurate measures of gap fraction as a function of zenith angle and azimuth. Advantages of an image approach include a permanent image which can be visually inspected and manually processed to accurately distinguish canopy elements, and which may be re-analyzed as methodologies develop.

Relatively inexpensive digital cameras are now available with a very large number of pixels that provide extremely high spatial resolution. Compared to photographic film, digital sensors have better radiometric quality and offer a number of practical advantages, not least among which is elimination of the expense and time involved with film processing and scanning, as well as the ability to preview images in the field to ensure image quality. Moreover, digital images have almost unlimited image treatment possibilities. Some caution should be used when employing typical "consumer-grade" digital cameras, as lens optics and dynamic range of such systems may introduce bias in canopy metric estimation (Frazer et al. 2001, Jonckheere et al. 2005).

An accurate estimate of LAI from digital hemispherical photography is best achieved when:

1. Exposure is controlled relative to above-canopy sky brightness (see Chen et al. 1991, Macfarlane et al. 2000, Zhang et al. 2005).
2. Images are corrected for the gamma function of the digital camera and its lens (see Leblanc 2005, Macfarlane et al. 2007).
3. The gap fraction distribution is corrected for foliage clumping (e.g. Chen and Cihlar (1995) and Lang and Xiang (1986) methods).

The LAI 2000 (LiCor, Lincoln, NB) plant canopy analyzer measures the gap fraction by comparing the radiation measured under the canopy at five zenith angles with simultaneous or near simultaneous above canopy incoming diffuse radiation, usually measured in a nearby opening. Where many under canopy measurements must be taken this can be a significant constraint if two instruments are not available. The LAI 2000 measures diffuse radiation simultaneously in five distinct angular bands, with central zenith angles of 7°, 23°, 53°, and 68°, by means of a fisheye light sensor. Chen et al. (1997) characterize the LiCor LAI 2000 as a more convenient and rapid version of hemispherical photography because image processing is not required. A built in optical filter blocks incoming radiation at wavelengths <490 nm to minimize canopy light scattering, maximizing contrast between the sky and canopy elements. The ratio of the two values gives the transmittance simultaneously for each sky sector and from which LAI is estimated. Calculations are performed automatically by internal software. Calculation assumptions include: foliage as an optically 'black body' which absorbs all light incident upon it; random distribution of canopy elements; canopy elements modeled using simple geometrical convex shapes; and, canopy elements assumed small relative to the angular bands of the instrument (Jonckheere et al. 2004). Advantages of this approach include its rapidity in continuous and homogeneous canopies. Disadvantages include the requirement of an above canopy reference; inaccuracy, mostly under-estimation, in clumped, discontinuous, and heterogeneous canopies; and, relatively coarse resolution, i.e. only five concentric rings using an immediate and automatic integration by the instrument in the field.

14.5.2.2 Canopy Transect Sampling

These approaches are typically line transect measurements of direct sunlight transmitted through canopy gaps yielding a profile of light that penetrates the canopy from which canopy gap size and distribution may be estimated. All require direct/full sunlight conditions for best operation. For gap fraction estimation, data collection over several hours is required to sample an adequate range of solar angles.

Demon (CSIRO, Canberra, Australia) is an optical instrument that is an extension of the point-quadrat method where direct beam radiation from the sun, measured through a directional narrow view angle, replaces the probe or plumb line (Lang 1990, Breda 2003). Filters are used to limit the spectrum of received light (≈430 nm) in order to minimize the effects of light scattering by foliage. A wide range of solar zenith angles is sampled by taking measurements, sampled via one or several line transects, over several hours and under bright sun-lit conditions to sample an adequate range of sun angles for gap fraction estimation. LAI is calculated from these data by model inversion and instrument specific averaging techniques (Dufrêne and Breda 1995).

AccuPAR and Sunfleck Ceptometer (Decagon Devices, Pullman, OR) are linear sensors that measure the sun fleck fraction or quantity of transmitted photsynthetically active radiation (PAR). These measurements, over a range of solar zenith angles and combined with assumptions about foliage distribution, are used with supplied software to estimate PAI. Limitations of these instruments include large variability between measurements necessitating substantial numbers of measurements in order to derive reliable results. The technique is not suitable in coniferous forests due to penumbral effects in the sun fleck fraction (Jonckheere et al. 2004).

TRAC (Tracing Radiation and Architecture of Canopies) is an optical instrument that measures the photosynthetic photon flux density (PPFD) through a canopy.

Using a solar beam as a probe, it records, by means of three photosensitive sensors, the transmitted direct light at a high frequency (Chen and Cihlar 1995). Sunflecks that are projected onto the sensor where canopy gaps occur in the sun's direction are proportional to the gap sizes being measured. By walking the instrument under a canopy at a fixed speed, its gap fraction and gap size distribution can be computed (Leblanc et al. 2002). The gap fraction measured by TRAC is with respect to the solar zenith angle at time of measurement. To effectively measure gap fraction, TRAC measurements at multiple solar zenith angles must be undertaken, such as over the course of half a day.

Using supplied software, the clumping index is found by comparing the measured gap size distribution with a theoretical gap size distribution associated with a canopy with randomly distributed foliage elements (Chen and Cihlar 1995). The clumping index obtained from TRAC can be used to convert L_e to LAI. A practical scheme for combining TRAC and LAI-2000 measurements in varied forest cover types is presented in Leblanc and Chen (2001).

14.6 Recommendations

14.6.1 Canopy Height

Ground-based measurements of canopy height are obtained from individual tree height measurements on a per plot basis. Total height and height to the base of the live crown should be recorded for each individual. Each individual should be tagged (see Chapter 4, Biomass) and its location from plot center (distance and azimuth) recorded. Laser rangefinders, sonic hypsometers, or clinometers are the recommended instrumentation.

14.6.2 Canopy Cover and Closure, Leaf Area Index

Digital hemispherical image analysis is recommended for the quantification of canopy cover, closure, and leaf (plant) area index. Variables measured using this approach include cover (using zenith angles $< 30^{o}$) and crown closure, as well as gap fraction, gap size distribution, and clumping index from which leaf area index may be derived.

Multiple images per plot should be obtained. Number of images is dependent on plot size. A grid or transect approach is recommended to capture spatial variability (e.g. Macfarlane et al. 2007, Jonckheere et al. 2005, Chen and Leblanc 2003).

The basic equation for obtaining LAI from optical measurements is (Chen 1996):

$$L = (1 - \alpha) L_{e\gamma E} / \Omega_E \tag{14.2}$$

where L denotes LAI; α is the woody-to-total plant area ratio; L_e is the effective LAI; γ_E is the needle-to-shoot area ratio; and Ω_E is the foliage clumping element index. A foliage element refers to either a conifer shoot or a broad leaf.

In optical gap fraction measurements, all objects above the ground, including both leaves and woody materials, affect L measurement. If no correction is made using the woody-to-total area ratio, i.e. $\alpha = 0$, then the value obtained is plant area index (PAI) rather than LAI. Since the principal interest is in green leaves only, effects of woody materials require removal by incorporating α. Sources for α in the literature include Chen (1996) and Gower et al. (1999). This ratio may be computed directly using the ratio between PAI derived from optical sampling and the maximum LAI value derived from direct measurement approaches such as litterfall.

As discussed previously, the starting point for optical estimates of LAI are measurements of radiation transmission from which gap fraction is estimated based on an assumption of a random leaf spatial distribution. As the distribution is often not random, the effective LAI, as estimated by the optical instruments, typically differs from the 'true' LAI. Non-randomness or 'clumping' may occur at two levels: at the level of the canopy element (leaf or shoot) and at scales larger than the canopy element. In conifer stands, needles are grouped first in shoots, hence shoots of conifers are treated as the foliage element and a correction for this leaf grouping is made using the needle-to-shoot area ratio, γ_E. For broad-leaved stands, individual leaves are the element and no shoot-level correction is necessary, i.e. $\gamma_E = 1$. Measurement of needle-to-shoot area ratio is highly labor intensive, so a recommendation is made here for use of published values where possible. Sources for needle-to-shoot area ratios for many forest species may be found in Chen (1996) and in Gower et al. (1999). Clumping at scales larger than the shoot is quantified using the element clumping index, Ω_E, which can be derived from optical measurements of canopy gap size distribution derived from hemispherical images.

Numerous software programs, some free and some commercially available, have been developed to process images and produce canopy metrics, e.g. WinSCANOPY (Regent Instruments, Qué., Canada), SOLARCALC (Chazdon and Field 1987), Winphot (ter Steege 1996), HemiView (Delta-T Device, Cambridge, UK), Gap Light Analyzer (Frazer et al. 1999), DHP (Leblanc et al. 2005), Hemisfer (Schleppi et al. 2007) and others.

Estimates of LAI derived from hemispherical optical sampling may be compared with direct estimates derived from litterfall sampling (see discussion above and Chapter 7, Litterfall Sampling).

Literature Cited

Adams J, Zuchowski L (1997) Impulse 200 and Impulse 200LR User's Manual 4th edition. Laser Technology, Inc., Englewood CA

Asner GP, Scurlock JMO, Hicke JA (2003) Global synthesis of leaf area index observations: implications for ecological and remote sensing studies. Global Ecology and Biogeography 12: 191–205

Avery TE, Burkhart HE (1983) Forest Measurements (3rd edition). McGraw-Hill, New York

Blair JB, Rabine DL, Hofton MA (1999) The Laser Vegetation Imaging Sensor (LVIS): A medium- altitude, digitisation-only, airborne laser altimeter for mapping vegetation and topography. ISPRS Journal of Photogrammetry and Remote Sensing. 54: 115–122

Breda NJJ (2003) Ground-based measurements of leaf area index: a review of methods, instruments and current controversies. Journal of Experimental Botany 54(392): 2404–2417

Bunnell FL, Vales DJ (1990) Comparison of methods for estimating forest overstory cover: differences among techniques. Canadian Journal of Forest Research 20: 101–107

Chazdon RL, Field C (1987) Photographic estimation of photosynthetically active radiation: evaluation of a computerized technique. Oecologia 73(4): 525–532

Chen JM (1996) Optically based methods for measuring seasonal variation in leaf area index of boreal conifer forests. Agricultural and Forest Meteorology 80: 135–163

Chen JM, Black TA (1992) Defining leaf area index for non-flat leaves. Plant, Cell, and Environment 15: 421–429

Chen JM, Cihlar J (1995) Quantifying the effect of canopy architecture on optical measurements of leaf area index using two gap size analysis methods. IEEE Transactions on Geoscience and Remote Sensing 33: 77–787

Chen JM, Rich PM, Gower ST, Norman JM, Plummer S (1997) Leaf area index of boreal forests: theory, techniques, and measurements. Journal of Geophysical Research 102 D24: 29, 429–29, 443

Chen JM, Leblanc SG (2003) Canopy structure measurements using optical instruments. In Fluxnet Canada Measurement Protocols 1.3. www.fluxnet-canada.ca

Chen JM, Black TA, Adams RS (1991) Evaluation of hemispherical photography for determining plant area index and geometry of a forest stand. Agricultural and Forest Meteorology 56: 129–143

Clark DB, Clark DA (2000) Landscape-scale variation in forest structure and biomass in a tropical rain forest. Forest Ecology and Management 137: 185–198

Cook JG, Stutzman, TW Bowers CW, Brenner KA, Irwin LL (1995) Spherical densiometers produce biased estimates of forest canopy cover. Wildlife Society Bulletin 23(4): 711–717

Donnelly DM, Johnson RR (1997) Westside Cascades Variant of the Forest Vegetation Simulator. WO-Forest Management Services Center. USDA Forest Service. Fort Collins, CO, 68 p

Drake JB (2001) Estimation of tropical forest aboveground biomass using large-footprint lidar. Ph.D. Thesis. University of Maryand, College Park, Maryland

Dubayah RO, Knox RG, Hofton MA, Blair JB, Drake JB (2000) Land surface characterization using lidar remote sensing In:. Hill MJ, Aspinall RJ (Eds.) Spatial information for land use management. Gordon and Breach Science Publishers, Australia. p.25–38

Dufrêne E, Breda N (1995) Estimation of deciduous LAI using direct and indirect methods. Oecologia 104: 156–162

Fassnacht KS, Gower ST, Norman JM, McMurtrie RE (1994) A comparison of optical and direct methods for estimating foliage surface area in forests. Agricultural and Forest Meteorology 71: 183–207

Fiala CS, Garman SL,Gray AN (2006) Comparison of five canopy cover estimation techniques in the western Oregon Cascades. Forest Ecology and Management. 232: 188–197

Franklin JF, Spies TA, Van Pelt R, Carey A, Thornburgh D, Berg DR, Lindenmayer D, Harmon M, Keeton WS, Shaw DC, Bible K, Chen J (2002) Disturbances and the structural development of natural forest ecosystems with some implications for silviculture. Forest Ecology and Management 155: 399–423

Frazer GW, Canham CD, Lertzman KP (1999) Gap Light Analyzer (GLA), Version 2.0: Imaging software to extract canopy structure and gap light transmission indices from true-color fisheye photographs. Copyright 1999: Simon Fraser University, Burnaby, BC, and the Institute of Ecosystem Studies, Millbrook, New York, (http://www.rem.sfu.ca/forestry/index.htm or http://www.ecostudies.org)

Frazer GW, Fourneir RA, Trofymow JA, Hall RJ (2001) A comparison of digital and film fisheye photography for analysis of forest canopy structure and gap light transmission. Agricultural and Forest Meteorology 109: 249–263

Gower ST, Kucharick CJ, Norman JM (1999) Direct and indirect estimation of leaf area index, fPAR, net primary production of terrestrial ecosystems. Remote Sensing of Environment 70: 29–51

Hansen M, DeFries R, Townshend JR, Carroll M, Dimiceli C, Sohlberg R (2003) Vegetation Continuous Fields MOD44B, 2001 Percent Tree Cover, Collection 3, University of Maryland, College Park, MD, 2001

Harding DJ, Lefsky MA, Parker GG, Blair JB (2001) Laser altimeter canopy height profiles. Methods and validation for closed-canopy, broadleaf forests. Remote Sensing of Environment 76: 283–297

Hemiview (Delta-T Devices) http://www.delta-t.co.uk/

Henning JG, Radtke PJ (2006) Detailed stem measurements of standing trees from ground-based scanning lidar. Forest Science 5291: 67–80

Hofton MA, Minster JB, Blair JB (2000) Decomposition of laser altimeter waveforms. IEEE Transactions on Geoscience and Remote Sensing 38: 1989–1996

Huang, C, Yang L, Wylie B, Homer C (2001) A strategy for estimating tree canopy density using Landsat 7 ETM + and high resolution images over large areas. USGS. EROS Data Center, Sioux Falls, South Dakota

Hyde P, Dubayah RO, Peterson BE, Blair JB, Hofton MA, Hunsaker C, Knox R, Walker W (2005) Mapping forest structure for wildlife habitat analysis using waveform lidar: Validation of montane ecosystems. Remote Sensing of Environment 96: 427–437

Jennings SB, Brown ND, Sheil D (1999) Assessing forest canopies and understory illumination: canopy closure, canopy cover and other measures. Forestry 72(1): 59–74

Jonckheere I, Fleck S, Nachaerts K, Muys B, Coppin P, Weiss M, Baret F (2004) Review of methods for in situ leaf area index determination. Part I. Theories, sensors and hemispherical photography. Agricultural and Forest Meteorology 121: 19–25

Jonckheere I, Nackaerts K, Muys B, Coppin P (2005) Assessment of automatic gap fraction estimation of forests from digital hemispherical photography. Agricultural and Forest Meteorology 132: 96–114

Kimes DS, Ranson KJ, Sun G, Blair JB (2006) Predicting lidar measured forest vertical structure from multi-angle spectral data. Remote Sensing of Environment 100: 503–511

Korhonen L, Korhonen KT, Rautiainen M, Stenburg P (2006) Estimation of forest canopy cover: a comparison of field measurement techniques. Silva Fennica 40(4): 577–588

Kucharik CJ, Norman JM, Gower ST (1998) Measurements of branch area and adjusting leaf area index indirect measurements. Agricultural and Forest Meteorology 91: 61–88

Kucharik CJ, Norman JM, Gower ST (1999) Characterization of radiation regimes in non-random forest canopies: theory, measurements, and a simplified modeling approach. Tree Physiology 19: 695–706

Kuusipalo J (1985) On the use of tree stand parameters in estimating light conditions below the canopy. Silva Fennica 19(2): 185–196

Lang ARG, Xiang Y (1986) Estimation of leaf area index from transmission of direct sunlight in discontinuous canopies. Agricultural and Forest Meteorology 37 (3): 229–243

Lang ARG (1990) An instrument for measuring canopy structure. Remote Sensing Reviews 5: 61–71

Leblanc SG, Chen JM (2001) A practical scheme for correcting multiple scattering effects of optical LAI measurements. Agriculture and Forest Meteorology 110:125–139

Leblanc SG, Chen JM, Kwong M (2002) Tracing Radiation and Architecture of Canopies TRAC Manual version 2.1.3. Natural Resources Canada

Leblanc SG, Chen JM, Fernandes R, Deering DW, Conley A (2005) Methodology comparison for canopy structure parameters extraction from digital hemispherical photography in boreal forests. Agricultural and Forest Meteorology 129: 187–207

Lefsky MA, Cohen WB, Parker GG, Harding DJ (2002) Lidar remote sensing for ecosystem studies. Bioscience 52(1): 19–30

Lefsky MA (1997) Application of lidar remote sensing to the estimation of forest canopy and stand structure. PhD. Dissertation. University of Virginia. Charlottesville, VA

LiCor http://www.licor.com/

Lovell JL, Jupp DLB, Culvenor DS, Coops MNC (2003) Using airborne and ground-based ranging lidar to measure canopy structure in Australian forests. Canadian Journal of Remote Sensing 29(5): 607–622

Macfarlane C, Coote M, White DA, Adams MA (2000) Agricultural and Forest Meteorology 107(2): 155–168

Macfarlane C, et al. (2007) Estimation of leaf area index in eucalypt forest using digital photography, Agricultural and Forest Meteorology. 143: 176–188

Norman JM, Campbell GS (1989) Canopy structure In: Percy RW, Ehleringer JR, Mooney HA, Rundel PW (Eds.) Plant physiological ecology: field methods and instrumentation. London: Chapman & Hall, pp. 301–325

Ollinger SV, Treuhaft RN, Braswell BH, Anderson JE, Martin ME, Smith ML (2007) The role of remote sensing in the study of terrestrial net primary production In: Fahey TJ, Knapp AK (Eds.). Principles and standards for measuring net primary production. Oxford University Press, Oxford, pp. 204–237

Parker GG (1995) Structure and microclimate of forest canopies. In: Lowman M, Nadkarni N (Eds.) Forest canopies: a review of research on a biological frontier. Academic Press, San Diego, CA, pp. 73–106

Parker GG, Brown MJ (2000) Forest canopy stratification-it it useful? The American Naturalist 155(4): 473–484

Parker GG, Harding DJ, Berger ML (2004) A portable lidar system for rapid determination of forest canopy structure. Journal of Applied Ecology 41: 755–767

Schleppi P, Sedivy I, Thimonier A (2007) Correcting non-linearities and slope effects in the estimation of the leaf area index of forests from hemispherical photographs. Agriculture and Forest Meteorology 144: 236–242 (http://www.wsl.ch/wald/dl/hemisfer/index-en.php)

Strickler GS (1959) Use of the densiometer to estimate density of forest canopy on permanent sample plots. USDA Forest Service. Portland, OR PNW. Note PNW-180. 5 p

ter Steege H (1996) WINPHOT 5.0: a programme to analyze vegetation indices, light and light quality from hemispherical photographs.Tropenbos Guyana Programme, Tropenbos, Guyana. Rep. 95–2. Wagner 1998 (http://www.bio.uu.nl/~herba/Guyana/winphot/wp_index.htm)

Treuhaft RN, Asner GP, Law BE, Van Tuyl S (2002) Forest leaf area density profiles from the quantitative fusion of radar and hyperspectral data. Journal of Geophysical Research-Atmospheres 107: 4568

Turner DP, Ollinger SV, Kimball JS (2004) Integrating remote sensing and ecosystem process models for landscape to regional scale analysis of the carbon cycle. BioScience 54: 573–584

Warren Wilson J. (1959) Analysis of the spatial distribution of foliage by two dimensional point quadrats. New Phytologist 58: 92–101

Welles JM (1990) Some indirect methods of estimating canopy structure. Remote Sensing Reviews 5: 31–43

Welles JM, Cohen S (1996) Canopy structure measurement by gap fraction analysis using commercial instrumentation. Journal of Experimental Botany 302: 1335–1342

WinsCanopy http://www.regentinstruments.com/products/Scanopy/Scanopy.html

Zhang YQ, Chen JM, Miller JR (2005) Determining digital hemispherical photograph exposure for leaf area index estimation. Agricultural and Forest Meteorology 133(1–4): 166–181

Chapter 15
Estimation of Forest Canopy Nitrogen Concentration

Marie-Louise Smith, David Y. Hollinger, and Scott Ollinger

Abstract The ability to detect patterns of carbon assimilation by vegetation is a key component of the North American Carbon Program. Because photosynthetic potential is strongly related to biochemical constituents such as nitrogen and chlorophyll concentrations in foliage, the ability to incorporate canopy chemistry into landscape- to regional-scale carbon cycling research represents an important contribution to NACP research goals. Here, we summarize the functional basis for using foliar N as a scalar of C uptake and recommend a practical, field-based approach for canopy N measurement that can be efficiently scaled using currently available remote sensing technology and ecosystem process modeling.

Keywords Forest productivity, light use efficiency, nitrogen, photosynthetic capacity, remote sensing, scaling

15.1 Introduction

The positive relationship between leaf nitrogen and photosynthetic capacity (e.g. Mooney and Gulmon 1979, Field and Mooney 1986, Reich et al. 1999) is one of the most robust results of modern ecophysiology and is at the core of a number of canopy productivity models (e.g. Aber et al. 1996). This relationship has as its

M.-L. Smith
US Forest Service, Northern Research Station, Durham, NH, Current address: US Forest Service, Legislative Affairs, 201 14th Street, SW, Washington, DC 20250-1130
E-mail: marielouisesmith@fs.fed.us

D.Y. Hollinger
US Forest Service, Northern Research Station, 271 Mast Road, Durham, NH 03824
E-mail: dhollinger@fs.fed.us

S. Ollinger
Complex Systems Research Center, University of New Hampshire, 56 College Road, Durham, NH 03824
E-mail: scott.ollinger@unh.edu

C.M. Hoover (ed.) *Field Measurements for Forest Carbon Monitoring*,

basis the fact that foliar nitrogen is found primarily in cellular proteins and that the principal carboxylating enzyme, ribulose-bisphosphate carboxylase-oxygenase (Rubisco) makes up a majority of total leaf protein (Björkman 1968).

Additional evidence supporting canopy N concentration as a scalar for C uptake comes from both theoretical and empirical studies. An important theoretical breakthrough was made by Sellers et al. (1992) when they related canopy nitrogen distribution to the canopy light environment and then to productivity. Because nitrogen is generally a scarce resource that limits plant carbon gain, it has been argued that natural selection should favor individuals that allocate N in an efficient manner through the canopy (Field 1983, 1991, Hirose et al. 1987, Hollinger 1989). Sellers et al. (1992) extended these arguments using a biochemically based model of leaf photosynthesis, coupled to a radiative transfer model, to show that canopy C uptake is maximized when N is allocated optimally within a plant canopy in proportion to the availability of photosynthetically active radiation.

Field measurements in tree canopies suggest that these patterns are a robust and general result (e.g. Hollinger 1999, Dang et al. 1997, Bond et al. 1999, Meir et al. 2002, Green et al. 2003). An important consequence of this relationship is that if the leaf nitrogen level is known at the top of the canopy, given a canopy light transmission model, whole canopy photosynthesis can be readily calculated. According to Sellers et al. (1992), canopies should allocate N such that for a specific light regime operating over a period T,

$$U_1 \int_0^T A dt + U_2 \int_0^T \frac{A}{N} dt \tag{15.1}$$

is maximized for all L, where L is the cumulative leaf area index, A is photosynthesis, and U_1 and U_2 are cost-benefit weighting factors, largely determined by the local availability of nitrogen. The first term relates to maximizing photosynthesis and the second term to maximizing efficiency of N use. Importantly, Sellers et al. (1992) findings are only valid for the case where $U_1 >> U_2$ and do not hold in very low N availability environments. A consequence of this is that we would expect the N: NPP relationship to break down at low nutrient sites such as bogs.

The solution of Sellers et al. (1992) takes the form of:

$$A_N = A_{n0} \Pi \tag{15.2}$$

where A_N is canopy photosynthesis produced by a certain amount of N, and A_{n0} and Π represent the separate effects of leaf biochemistry and canopy structure, respectively. A_{n0} can be thought of as a "single leaf" solution where the N at the top of the canopy sets maximum photosynthetic capacity and Π is a scaling factor to relate the "top" leaf performance to canopy performance. For a simple, exponential model of light attenuation,

$$\Pi = \int_0^{LT} \overline{f(L)}\, dL = \left[\frac{1 - e^{-kLT}}{\bar{k}}\right] = \frac{\overline{FPAR}}{\bar{k}} \tag{15.3}$$

where k is the canopy light extinction coefficient and FPAR is the fraction of absorbed photosynthetically active radiation and the overbars denote a radiation-weighted mean value.

In other words, whole-canopy photosynthesis (GPP) is a simple function of leaf N at the top of the canopy and APAR. This result provides important simplifications and constraints for models of land surface CO_2 exchange. Data from several stand-level studies support these notions, having demonstrated significant positive relationships between mass-based foliar N concentrations and maximum net photosynthesis or A_{max} (Reich et al. 1995, 1999), and between aboveground net primary production (ANPP) and mass-based foliar N concentrations averaged over the entire canopy (Comeau and Kimmins 1985, Smith et al. 2002). A study by Green et al. (2003) highlighted the prominent role of N in explaining canopy light use efficiency across a broad array of C3 plants, providing additional evidence that the same factors responsible for the efficiency of leaf-level photosynthesis have strong potential to be directly scaled to whole plant canopies.

A promising extension of these results is the development of methods for quantifying canopy nitrogen concentrations that can substantially improve the accuracy of model-based production estimates derived using remote sensing (Ollinger and Smith 2005). Methods for estimating canopy N using high spectral resolution remote sensing have been in existence for some time (Matson et al. 1994, Zagloski et al. 1996, Martin and Aber 1997), and have been applied with a high degree of accuracy using high spectral resolution data from NASA's airborne imaging spectrometer, AVIRIS (Martin and Aber 1997, Smith et al. 2002, Asner and Vitousek 2005). A robust capacity for remote detection of canopy N from a space-based platform (Hyperion) has also recently been demonstrated within a number of forest biomes (Smith et al. 2003, Townsend et al. 2003, Coops et al. 2003, Martin et al. 2008). Regional- to continental-scale gradients in foliar N have been observed through analysis of field measurements and through combined field and hyperspectral remote sensing campaigns across multiple biomes (e.g. Yin 1992, Aber et al. 2003, McNeil 2006, Martin et al. in press). Methods for canopy N detection from broader-scale instruments such as MODIS are beginning to be developed (Potter et al. 2007, Ollinger et al. 2007).

15.2 Field Measurement

The ability to efficiently scale leaf-level traits to whole forest canopies enhances our ability to examine key relationships associated with these traits at various levels from the leaf to the forest stand and, with remote sensing technologies, to larger landscapes. Field-based estimates of forest canopy chemistry are based on leaf-level measurements of mass and chemistry combined with estimates of the fractional distribution of species' leaf area in a forest canopy. The methods outlined below are largely drawn from Smith and Martin (2001) and Smith et al. (2002).

15.2.1 Green-Leaf Chemistry and Leaf Mass Per Unit Area (LMA)

On a per plot basis, two essential measurements are required: an estimate of the mean N concentration (mg/g) for each species on each plot combined with a mean value of leaf mass per area (g/cm^2, LMA) for each species.

A representative number of leaf samples are needed to derive a robust mean leaf-level N. In order to determine growing season foliar chemistry, mid-growing season green leaf samples should be collected on each study plot. On each plot, all canopy species should be identified and, depending on canopy composition, between two and seven trees per species selected for green leaf collection. Leaves can be quite efficiently collected in moderately tall canopies (≤35 m) by shooting small branches from the canopy using shotguns (12 gauge, #3 or #4 steel shot). Each sample should consist of leaves collected from several places in the canopy. Leaf-level N can be determined by various methods including CHN analysis, laboratory spectroscopy (Bolster et al. 1996), or wet chemistry. Additional leaf samples should be collected from the upper canopy of each dominant or co-dominant species on each plot in order to make a determination of LMA for each species on a per plot basis (Chapter 14, Canopy Structure section on considerations for LMA sampling and measurement).

15.2.2 Scaling Leaf Level N to Whole Canopies

As N concentration on a mass basis does not vary significantly in relation to vertical canopy gradients (Ellsworth and Reich 1993, O'Neill et al. 2002), canopy-level nitrogen concentration (mg/g) on a per plot basis can be simply calculated as the sum of the mean of foliar N concentrations (mg/g) for individual species in each stand, weighted by fraction of canopy foliar mass per species.

Determination of canopy species fraction by mass can be accomplished variously. One method employs litterfall collection where canopy species fraction by mass is calculated directly from dried, sorted litterfall data. This method is most appropriate in deciduous stands where litterfall is equal to canopy production and complete leaf senescence occurs at the end of the growing season. In mixed or pure conifer stands this method is problematic and not recommended. Needle retention varies among species and from year to year based on climatic conditions and, absent a record of litter collection over multiple seasons, both total and species fraction of canopy mass will tend to be underestimated (see Chapter 14, Canopy Structure section).

The alternative and recommended method, validated against litterfall data from both deciduous and mixed stands, employs a camera-based point quadrat sampling technique to estimate species fraction of leaf area (Aber 1979a, b) and species LMA measurements (Smith and Martin 2001). A 35 mm camera with a 135 mm telephoto lens serves as the sampling device (Fig. 15.1).

The focal plane of the lens is calibrated to distance in meters to allow use as a range finder, and a grid of 15 points is marked on the camera's viewing screen (such

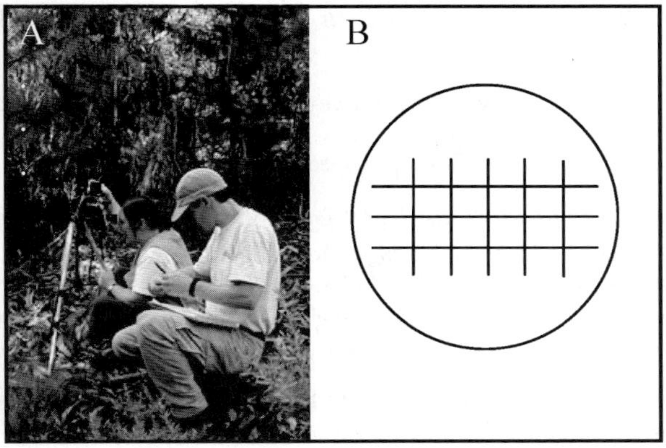

Fig. 15.1 Camera-based point quadrat sampling to estimate species' fraction of leaf area in a forest stand. A 35 mm camera is used as the sampling device (A). The focal plane of the lens, calibrated to distance, is used as a range finder and the camera's viewing screen is used as the sampling grid (B)

grids are commercially available). In each sample plot the camera, mounted on 1 m tall tripod, is directed upward towards the canopy and leveled. The species and height of the lowest leaf covering each grid point is determined by focusing the lens and recording the calibrated distance. In each sample plot, 15 grid-point observations are taken at nine sample points for a total of 135 observations per plot. Although not an accurate estimator of total leaf area, this method has been demonstrated to be an accurate means of determining the relative distribution or fraction of leaf area by height (MacArthur and Horn 1969, Aber 1979a, b) and by species (Parker et al. 1989) in a forested canopy.

Using the equation of Aber (1979a) LAI is calculated above a series of heights in the canopy (at 2 m increments from 2–38 m):

$$LAI_h = \ln\left(\frac{N_h + S}{S}\right) \tag{15.4}$$

where:

h is the canopy height,
LAI is the leaf area index above height h,
N is the number of leaf intercepts above height h,
S is the total number of sky point intercepts.

From this calculation LAI is determined within each of the 2 m vertical increments and the fraction of each species within each increment. The LAI attributed to each species is summed through the vertical profile and divided by the total canopy LAI to derive the fraction of total canopy LAI by species. Fraction of species by leaf area is converted to fraction by weight by multiplying area fraction by measured specific leaf weight of each species and deriving a new fraction by weight for each species on sample plots.

Literature Cited

Aber JD (1979a) A method for estimating foliage-height profiles in broad-leaved forests. Journal of Ecology 67:35–40

Aber JD (1979b) Foliage-height profiles and succession in northern hardwood forests. Ecology 60:18–23

Aber JD, Goodale CL, Ollinger SV, Smith ML, Magill AH, Martin ME, Hallett RA, Stoddard JL and NERC Participants (2003) Is nitrogen deposition altering the nitrogen status of northeastern forests? Bioscience 53(4):375–389

Aber JD, Reich PB, Goulden ML (1996) Extrapolating leaf CO_2 exchange to the canopy: a generalized model of forest photosynthesis validated by eddy correlation. Oecologia 106:257–265

Asner GP, Vitousek PM (2005) Remote detection of biological invasion and biochemical change. Proceedings of the National Academy of Sciences 102 (12):4383–4386

Bolster K, Martin ME, Aber JD (1996) Determination of carbon fraction and nitrogen concentration in tree foliage by near infrared reflectance: a comparison of statistical methods. Canadian Journal of Forest Research 26:590–600

Bond BJ, Farnsworth, BT, Coulombe RA, Winner WE (1999) Foliage physiology and biochemistry in response to light gradients in conifers with varying shade tolerance. Oecologia 120:183–92

Björkman O (1968) Carboxydismutase activity in shade-adapted and sun-adapted species of higher plants. Physiol. Plantarum 21:1–10

Comeau PG, JP Kimmins (1985) The relationship between net primary production and foliage nitrogen content, and its application in the modelling of forest ecosystems: a study of Lodgepole pine. Fugimori T (ed.). International Union of Forest Research Organizations

Coops NC, Stone C, Culvenor DS, Chisholm LA, Merton RN (2003) Chlorophyll content in eucalypt vegetation at the leaf and canopy scales as derived from high resolution spectral data. Tree Physiology 23(1):23–31

Dang QL, Margolis HA, Sy M, Coyea MR, Collatz GJ, Walthall CL (1997) Profiles of photosynthetically active radiation, nitrogen and photosynthetic capacity in the boreal forest: Implications for scaling from leaf to canopy. Journal of Geophysical Research – Atmospheres 102:28845–28859

Ellsworth DS, Reich PB (1993) Canopy structure and vertical patterns of photosynthesis and related leaf traits in a deciduous forest. Oecologia 96:169–178

Field C (1983) Allocating leaf nitrogen for the maximization of carbon gain: Leaf age as a control on the allocation program. Oecologia 56:341–347

Field C (1991) Ecological scaling of carbon gain to stress and resource availability. In: Mooney HA, Winner WE, Pell EJ (Eds.) Response of plants to multiple stresses. Academic Press, New York, pp. 35–65

Field C, Mooney HA (1986) The photosynthesis-nitrogen relationship in wild plants. In: Givnish TJ (ed.).On the economy of plant form and function. Proceedings of the Sixth Maria Moors Cabot symposium, evolutionary constraints on primary productivity: adaptive patterns of energy capture in plants, Harvard Forest, August 1983. Cambridge University Press, Cambridge, pp. 25–55

Green DS, Erickson JE, Kruger EL (2003) Foliar morphology and canopy nitrogen as predictors of light-use efficiency in terrestrial vegetation. Agricultural and Forest Meteorology 3097:1–9

Hirose T, Werger MJA, Rheenen JWA (1989) Canopy development and leaf nitrogen distribution in a stand of *Carex acutiformis.* Ecology 70:1610–1618

Hollinger DY (1989) Canopy organization and foliage photosynthetic capacity in a broad-leaved evergreen montane forest. Functional Ecology 3:53–62

Hollinger DY (1999) Optimality and nitrogen allocation in a tree canopy. Tree Physiology 16: 627–34

MacArthur RH, Horn HS (1969) Foliage profile by vertical measurements. Ecology 50:802–804

Martin ME, Aber JD (1997) High spectral resolution remote sensing of forest canopy lignin, nitrogen, and ecosystem processes. Ecological Applications 7:431–443

Martin ME, Plourde L, Smith ML, Ollinger SV. (2008) A generalizable method for remote sensing of canopy nitrogen across a wide range of forest ecosystems. Remote Sensing of Environment (in press)

McNeil BE (2006) Spatial variability of foliar nitrogen in the Adirondack Park, New York, Ph.D. Dissertation, Department of Geography, Syracuse University, 223 p

Matson PA, Johnson L, Billow C, Miller J, Pu R (1994) Seasonal patterns and remote spectral estimation of canopy chemistry across the Oregon transect. Ecological Applications 4:280–298

Meir P, Kruijt B, Broadmeadow M, Barbosa E, Kull O, Carswell F, Nobre A, Jarvis PG (2002) Acclimation of photosynthetic capacity to irradiance in tree canopies in relation to leaf nitrogen concentration and leaf mass per unit area. Plant, Cell and Environment 25:343–357

Mooney HA, Gulmon SL (1979) Environmental and evolutionary constraints on photosynthetic characteristics of higher plants. In: Solbrig OT, Jain S, Johnson GB, Raven PH (Eds.) Topics in plant population biology. Columbia University Press, New York, pp. 316–337

Ollinger SV, Smith ML (2005) Net primary production and canopy nitrogen in a temperate forest landscape: an analysis using imaging spectrometry, modeling, and field data. Ecosystems 8:760–778

Ollinger SV, Treuhaft RN, Braswell BH, Anderson JE, Martin ME, Smith ML (2007) The role of remote sensing in the study of terrestrial net primary production. In: Fahey TJ, Knapp AK (eds.). Principles and standards for measuring net primary production, Oxford University Press, Oxford, pp. 204–237

O'Neill AL, Kupiec JA, Curran PJ (2002) Biochemical and reflectance variation throughout a Sitka spruce canopy. Remote Sensing of Environment 80:34–142

Parker GG, O'Neil JP, Higman D (1989) Vertical profile and canopy organization in a mixed deciduous forest. Vegetatio 85:1–11

Potter C, Gross P, Genovese V, Smith ML (2007) Net primary productivity of forest stands in New Hampshire estimated from Landsat and MODIS satellite data. Carbon Balance and Management 2:9

Reich P, Kloeppel B, Ellsworth D, Walters M (1995) Different photosynthesis-nitrogen relations in deciduous hardwood and evergreen coniferous tree species. Oecologia 104:24–30

Reich PB, Ellsworth DS, Walters MB, Vose JM, Gresham C, Volin JC, Bowman WD (1999) Generality of leaf traits relationships: a test across six biomes. Ecology 80:1955–1969

Sellers PJ, Berry JA, Collatz GJ, Field CB, Hall FG (1992) Canopy reflectance, photosynthesis, and transpiration. III. A reanalysis using improved leaf models and a new canopy integration scheme. Remote Sensing of Environment 42:187–216

Smith ML, Martin ME (2001) A plot-based method for rapid estimation of forest canopy chemistry. Canadian Journal of Forest Research 31:549–555

Smith ML, Martin ME, Ollinger SV, Plourde L (2003) Analysis of hyperspectral data for estimation of temperate forest canopy nitrogen concentration: comparison between an airborne (AVIRIS) and a spaceborne (HYPERION) sensor. IEEE Transactions on Geosciences and Remote Sensing 41:1332–1337

Smith ML, Ollinger SV, Martin ME, Aber JD, Hallett RA, Goodale CL (2002) Direct estimation of aboveground forest productivity through hyperspectral remote sensing of canopy nitrogen. Ecological Applications 12:1286–1302

Townsend PA, Foster JR, Chastain RA Jr, Currie WS (2003) Imaging spectroscopy and canopy nitrogen: application to the forests of the central Appalachian Mountains using Hyperion and AVIRIS. IEEE Transactions on Geoscience and Remote Sensing 41:1347–1354

Yin X (1992) Empirical relationships between temperature and nitrogen availability across North American forests. Canadian Journal of Forest Research 22:707–712

Zagloski F, Pinel V, Romier J, Alcayde D, Gastellu-Etchegorry JP, Giodano G, Marty G, Mougin E (1996) Forest canopy chemistry with high spectral resolution remote sensing. International Journal of Remote Sensing 17:1107–1128

Part VI
Lessons from the Past and Opportunities in the Future

Chapter 16
Integrating Field Measurements with Flux Tower and Remote Sensing Data

Kenneth J. Davis

Abstract Landscape-scale forest carbon cycle measurements are needed to map forest carbon cycles across space and time, and to identify the processes governing the associated fluxes. Landscape-scale forest carbon cycle measurements are intended to be integrated with flux tower and remote sensing measurements into a diagnosis of forest-atmosphere carbon fluxes referred to in this chapter as flux tower upscaling. Flux tower upscaling is intended to fill gaps in our current ability to diagnose the terrestrial carbon cycle. More complete diagnoses of contemporary fluxes will enable more accurate prediction of future fluxes. Flux tower upscaling is a portion of a broader North American sampling strategy that also includes biomass and atmospheric inventory measurements. Flux tower upscaling has the potential to yield more mechanistic understanding of the terrestrial carbon cycle than inventory approaches, understanding that is essential for developing prognostic skill. Initial demonstrations of flux upscaling in forest ecosystems exist, as well as initial comparisons to inventory methods. Research concerning the design of the landscape-scale forest carbon cycle measurement network required for effective flux tower upscaling is essential, as is research concerning integration of flux tower upscaling with inventory measurements, manipulative experiments, and integrative earth-system studies. This research should include integrative field experiments, and efforts to improve and create key observational methods. Centralized data management and coherent, long-term support for the network observations found essential to this effort are critical for the success of the flux tower upscaling effort, and the North American Carbon Program in general.

Keywords Biomass inventory, flux tower upscaling, forest carbon cycle, network design, North American Carbon Program

---

K.J. Davis
Department of Meteorology, The Pennsylvania State University, 512 Walker Building. University Park, PA 16802
E-mail: davis@meteo.psu.edu

C.M. Hoover (ed.) *Field Measurements for Forest Carbon Monitoring*,

16.1 Introduction

Understanding the earth's terrestrial carbon cycle is a challenging fundamental research problem. In addition, societal interest in managing climate change is motivating intense applied study of this topic. Societal applications include diagnosing fluxes for regulatory purposes and predicting future terrestrial sources and sinks of CO_2 to guide carbon management policy. Diagnosing fluxes at regional to continental scales is essential to evaluating with accuracy our collective contributions to the earth's greenhouse gas burden, yet our current ability to diagnose fluxes at these spatial scales is limited (e.g. Pacala et al. 2001, Janssens et al. 2003). Predicting future terrestrial carbon fluxes is essential to developing optimal climate management strategies. Current predictions of the coupled climate-terrestrial carbon cycle system are highly uncertain (Friedlingstein et al. 2006). Prediction requires mechanistic understanding that may not be essential to diagnosis. Diagnosis and prediction of the terrestrial carbon cycle at regional to continental scales are two central goals of current terrestrial carbon cycle research.

A necessary objective of this community research effort is the development of an observational system that characterizes terrestrial carbon stocks and fluxes across North America with the accuracy, precision, and spatial and temporal resolution sufficient to develop a mechanistic understanding of the terrestrial carbon balance. Such a system would satisfy the need to evaluate regulatory strategies with regional flux measurements. Such a system should also be maintained long enough and include enough mechanistic detail to observe the response of the terrestrial carbon cycle to the driving factors whose impacts we would like to predict (e.g. climate change, land use change, nutrient cycling, disturbance), and to attribute observed changes to those driving factors. This observational system, combined with manipulative ecosystem experiments that attempt to simulate future environments, would provide a sound basis on which we could develop our capacity to predict the future terrestrial carbon cycle.

This observational system, however, does not yet exist. A fundamental methodological problem facing the carbon cycle science research community is the fact that no single measurement technique is suited to capture the full range of spatial and temporal scales that need to be characterized. In particular the community lacks methods suited to address intermediate spatial scales (Fig. 16.1). Such methodological development is needed if the research community is to succeed in developing the ability to: (1) observe the terrestrial carbon cycle of North America (diagnose), (2) understand the mechanisms governing this cycle (attribute) and (3) describe how the terrestrial carbon cycle of North America is going to evolve with time (predict).

The North American Carbon Program (NACP, Wofsy and Harriss 2002) and the NACP science implementation strategy (Denning et al. 2005) proposed a design for this observing system. It is important to recall that this observational design remains a working hypothesis. The system proposed by Denning et al. (2005) includes a "Tier 2" set of measurements intermediate between flux towers ("Tier 1", dense in information content, sparse in spatial sampling) and inventories and remote sensing

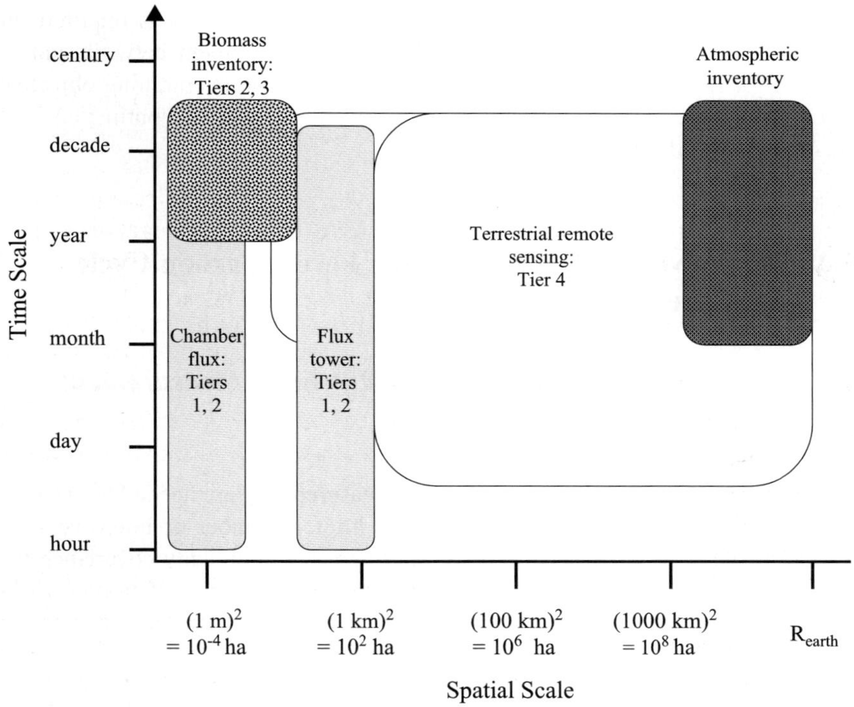

Fig. 16.1 Intrinsic spatial and temporal scales of the measurement techniques utilized in terrestrial carbon cycle studies, and their place within the tier structure of the North American Carbon Program. The minimum temporal scale is a function of the temporal resolution of the method. The maximum temporal scale grows as record lengths increase. Networks of mixed tiers of measurements can be assembled in various ways to encompass larger spatial scales. The merging of flux and inventory measurements with remote sensing and terrestrial carbon cycle models to accomplish this task, a method referred to as flux tower upscaling, is the focus of this chapter

(less information dense, more comprehensive spatial coverage, Tiers 3 and 4, respectively). This chapter focuses on the methods and challenges involved in integrating these intermediate intensity measurements with flux tower and remote sensing data into a spatially complete "bottom-up" flux estimate. This integration will be referred to as flux tower upscaling. As proposed by Denning et al. (2005), flux tower upscaling will provide regional to continental scale flux estimates that complement quantification of the terrestrial carbon cycle based on biomass and atmospheric inventory measurements (Fig. 16.1 and Denning et al. 2005).

Despite the abundance of well-reasoned measurement and sampling strategies presented in this manual, the precise design of Tier 2 measurements (referred to in this manual as landscape-scale forest carbon cycle measurements) remains a research topic. At several points within this manual authors rightfully note that the sampling protocol required to achieve a given degree of accuracy in characterizing particular landscape-scale carbon stocks or fluxes is not well known. The sampling strategy is even less certain when considering how best to integrate multiple measurements across the NACP measurement tiers. A goal-oriented definition is required; the design of the landscape-scale forest carbon cycle measurement network and the strategies for flux tower upscaling must be dictated by the overarching objectives of developing the ability to diagnose and predict regional to continental scale terrestrial carbon fluxes.

16.2 Objectives of Landscape-Scale Forest Carbon Cycle Measurements

16.2.1 Expand the Interpretation of Intensive Measurement Sites Across Space

Landscape-scale forest carbon cycle (Tier 2) measurements are needed to extend in space the understanding gained from a more limited number of intensive study sites, designated Tier 1 sites, conceptualized to consist of an eddy-covariance flux tower augmented by a suite of ongoing complementary measures of carbon stocks, carbon fluxes and relevant biogeochemical and ecological status, and currently realized in measurement collectives such as AmeriFlux.

Hollinger's opening chapter of this manual describes the spatial characteristics of measurements needed to satisfy this objective. He notes that this tier of measurements must characterize the spatial distribution of carbon stocks and fluxes across a greater area than is sampled by a single intensive site (e.g. a flux tower and its associated flux footprint, Horst and Weil 1992) in order to evaluate how the intensive site is representative of the greater landscape within which it is located. In addition, the landscape-scale forest carbon cycle measurements must also encompass one or more satellite pixels, in order to enable the close integration of flux tower and remote sensing observations into continental-scale carbon flux estimates (Fig. 16.1).

In order to satisfy this objective, landscape-scale forest carbon cycle (Tier 2) measurements *must* be coordinated with intensive (Tier 1) measurement sites.

16.2.2 Enable Methodological Overlap in Regional Flux Estimates

The objective of this integration of flux towers and remote sensing is not just to create a single continental-scale carbon flux estimate, but also to enable comparisons of multiple independent measures of the regional to continental scale terrestrial carbon cycle. Diagnosis and prediction of regional to continental carbon fluxes requires the development of accurate flux measurement methods applicable to these scales. Demonstration of accurate measurements requires that multiple independent measures are employed simultaneously. Though there is no consensus regarding how to integrate vs. compare measurements at this time, continental-scale methodological comparison efforts to date (Pacala et al. 2001, Janssens et al. 2003) have grouped existing methods into three categories: atmospheric inventory measurements, biomass inventory measurements, and flux upscaling measurements (see Fig. 16.1). In the framework described by Hollinger in Chapter 1, landscape-scale carbon cycle measurements within the NACP observing plan are used to link remote sensing and intensive measurements into one "flux measurement" methodology for deriving regional to continental fluxes, one of two so-called "bottom-up" approaches described in NACP planning documents. This approach can be compared independently to both biomass (e.g. Houghton 2003, Jain and Yang 2005) and atmospheric (e.g. Gurney et al. 2002) inventory measurements.

As methodological consensus emerges, it is likely that a streamlined observational system integrating these three basic approaches will be developed. At present the community is engaged in the methodological and mechanistic experiments intended to identify the essential elements of this observational network.

16.2.3 Integrate Across Components of the Forest Carbon Cycle

Landscape-scale forest carbon cycle (Tier 2) measurements must integrate across component fluxes, stocks and processes relevant to the terrestrial carbon cycle in addition to providing a link across temporal and spatial scales. Figure 16.1 and the related discussion of integration across spatial and temporal scales do not make clear the fact that the observational methods being discussed (remote sensing, flux towers, landscape-scale forest carbon cycle inventory and component flux measurements) not only observe different spatial and temporal scales, but also different components of the forest carbon cycle. Figure 16.2 illustrates the components of the forest carbon cycle typically captured by the various observational approaches utilized by the research community. Landscape-scale measurements ideally provide the observations that are needed to integrate flux measurements and remote sensing not just across spatial scales, but also across component fluxes of the forest carbon cycle. This problem of integrating component fluxes within the terrestrial carbon cycle is at least as challenging an aspect of the design and implementation of this tier of the NACP science plan as the issue of integration across spatial scales. Characterization of the below-ground carbon cycle is probably the most difficult facet of integrating across components of the carbon cycle, since it is not readily detected via remote sensing. It is unlikely that direct observation of all components of the terrestrial carbon cycle at landscape scales can be implemented; terrestrial ecosystem models and mechanistic understanding of the cycles must be brought to bear to construct a sufficiently complete understanding. This objective is reflected in Hollinger's call in Chapter 1 for landscape-scale measurements to provide the ecological data needed to parameterize models of the terrestrial carbon cycle.

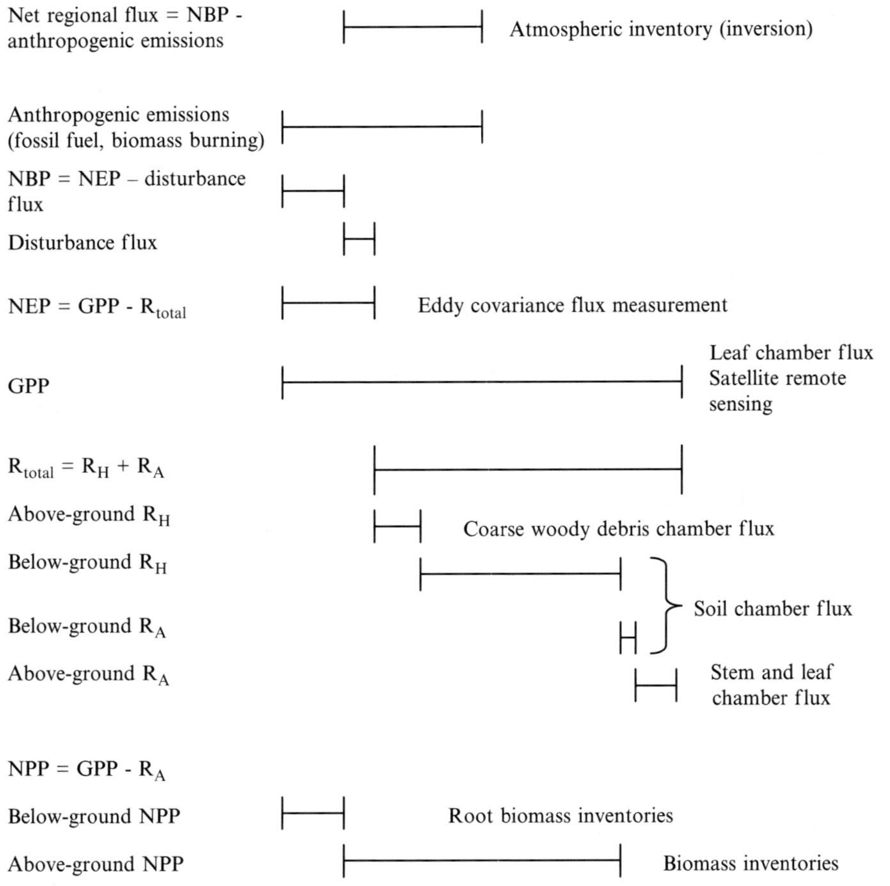

Fig. 16.2 Schematic of the relationship between component fluxes of the forest carbon cycle and the observational methods outlined in Fig. 16.1. Magnitudes of the bars, while not precise, are intended to represent plausible magnitudes of the fluxes. Component fluxes are named on the left. Methods for observing those fluxes are on the right. Acronyms used are NBP (net biome productivity), NEP (net ecosystem productivity), R (respiration), R_A (autotrophic respiration), and R_H (heterotrophic respiration)

16.2.4 Enable Mechanistic Understanding of the Forest Carbon Cycle

The final objective that we wish to address by the integration of flux towers, remote sensing and landscape scale forest carbon cycle measurements is improving the mechanistic understanding needed to predict the future carbon cycle in forests. Neither flux towers nor remote sensing observations directly evaluate the processes that govern the observed fluxes. Such interpretation must be done with a combination of process-oriented observations and models of the terrestrial carbon cycle.

The landscape-scale forest carbon cycle measurements discussed in this manual must provide the additional information needed to identify the mechanisms governing the observed fluxes.

The governing mechanisms that are clearly relevant to the forest carbon cycle on time scales from hours to centuries include climate (light, temperature and moisture), nutrient cycling, disturbance (fire, disease, insects, wind throw and human land management) and succession. Without a landscape-scale understanding of these governing factors, and mechanistic models that incorporate them accurately, the integration of fluxes, remote sensing and inventory measurements (both atmospheric and biomass) into a predictive understanding of the terrestrial carbon cycle is not possible. While these landscape-scale carbon cycle measurements are not intended to enable prediction of all these governing factors (e.g. not intended to provide all data needed to drive prognostic models of climate change or human land use), observations should be designed to permit the attribution of the response of the forest carbon cycle to these mechanisms (e.g. identify changing climate or human land use as a cause of changes in the forest carbon cycle). For example, measurements across gradients in land use history allow us to attribute spatial variability in landscape-scale carbon cycling to disturbance, allowing predictions to be made of the response of the terrestrial carbon cycle to land use change scenarios.

16.3 Progress Toward Regional to Continental Flux Tower Upscaling

Inventories, biomass and atmospheric, have been used most often to diagnose regional (most often biomass inventories) to continental (more accessible to atmospheric inventories) carbon budgets. Pacala et al. (2001), for example, applied both methods to North America. Janssens et al. (2003) performed a similar study of the European carbon budget but added carbon dioxide flux measurements from a European flux tower network and satellite-based land surface classifications, presenting a simple example of the integration of flux towers and remote sensing. The NACP science plan calls for more sophisticated integration that includes the tier of measurements described in this manual to improve the accuracy of flux tower upscaling efforts.

16.3.1 Methodological Building Blocks

Most integrative efforts to date have combined only a subset of the proposed array of NACP "bottom-up" measurements, and often at single intensive measurement sites rather than over a spatial array intended to extend the flux tower measurements over a larger region. These studies are important building blocks towards achieving effective regional flux tower upscaling. This chapter is not intended as a literature review, but a brief review of some representative publications will help to build the context around the task of integration across space and methods.

A first step toward integration is comparison. Direct comparisons between flux tower measurements and many of the measurement methods described in this manual, including forest inventory measurements (Barford et al. 2001, Gough et al. 2008), chamber flux measurements (e.g. Goulden et al. 1996, LaVigne et al. 1997, Loescher et al. 2006, Tang et al. 2008) and remote sensing based estimates of gross primary production (GPP) (Heinsch et al. 2006) conducted at flux tower sites have made important progress in exploring how these measurement approaches can be integrated, and the sampling required to obtain comparable results. Flux tower sites have also hosted extensive efforts to compare satellite-derived measures of leaf area index, land cover type, NPP and GPP (Cohen et al. 2006, Turner et al. 2006) to ground-based measurements similar to those proposed in this manual, building knowledge of the methods needed to convert satellite radiances into quantitative understanding of the forest carbon cycle.

Observations alone are not sufficient to achieve the task of integration. The mechanistic understanding of the forest carbon cycle embedded in terrestrial carbon cycle models must be used to bridge the gaps in scales and processes (Figs. 16.1 and 16.2) that prevent direct observational upscaling to regional fluxes. Numerous publications present comparisons between observations (remote sensing, flux towers, ground-based forest carbon cycle measurements) and numerical models of the forest carbon cycle. The traditional approach has been to use measurements of forest biophysical state and processes (e.g. soil carbon pool sizes, stomatal conductance of leaves) to determine model parameters for given species or biomes, and compare the resulting model predictions driven by environmental conditions, such as climate, to independent flux observations such as eddy covariance, chambers, biomass accumulation rates and sap fluxes (e.g. Hanson et al. 2004, Thornton et al. 2002, Kucharik et al. 2006). Comparisons of this sort have yielded valuable understanding concerning the validity of the numerical models across a range of time scales and forest systems. This approach retains methodological independence between the process models and the data used to determine model parameters on one hand, and the flux observations on the other. It is often difficult, however, to isolate the model parameters or structures that are responsible for discrepancies between models and the independent flux observations. This approach has also been used as a route towards improvement of terrestrial carbon cycle models, but the integration of the information into a process model is done in an ad-hoc fashion.

The need to integrate these complex, multi-source data sets in a more methodical fashion has motivated increasing adoption of data assimilation methods. Measurements can be compared to terrestrial carbon cycle model output and model parameters adjusted so that the discrepancy between measurements and model output is minimized. In this way, large and complex data sets can be integrated via the models into a single interpretation of fluxes and their drivers, similar to long-standing practices in operational weather forecasting. Unlike weather forecasting, however, the processes governing the terrestrial carbon cycle are not as well understood, so at this time the greatest value in this approach for terrestrial carbon cycle studies lies in evaluation of the model parameters, model structures and observations needed to evaluate the models. Braswell et al. (2005), for example, show that a widely used terrestrial carbon cycle model is not structurally capable of capturing interannual variability observed in a long-term flux tower record. Knorr and Kattge (2005) illustrate how to compare the prior knowledge of model parameters derived from process measurements to whole-ecosystem flux measurements in a mathematically rigorous fashion, producing parameter probability distributions and identifying the model parameters for which the two sources of data provide consistent vs. conflicting information. These approaches show promise for guiding rigorous development of both models and observations, evaluating predictive capacity of carbon cycle models and integrating models and data needed to perform effective upscaling of fluxes across regions.

16.3.2 Existing Efforts Regarding Regional Flux Upscaling

A number of research efforts in North America, building on the foundation of studies outlined above, have begun to assemble elements of the regional flux upscaling approach called for in the NACP science implementation strategy. Turner et al. (2007) integrated remotely sensed forest disturbance data and a terrestrial carbon cycle model tuned to regional eddy covariance measurements into a multi-year estimate of terrestrial carbon balance for the state of Oregon. A regional study in northern Wisconsin employed two approaches: a regional network of flux towers (Desai et al. 2008), and spatial decomposition of tall-tower flux measurements (Wang et al. 2006) merged with remotely sensed land cover and simple carbon cycle models, to create two largely independent regional carbon exchange estimates. Desai et al. (2008) found fairly good agreement between the two products. An experiment at a boreal forest site in Manitoba brought together a regional network of flux towers (Goulden et al. 2006) and complementary forest carbon stock and chamber flux measurements (Bond-Lamberty et al. 2004) spanning a chronosequence of seven stand ages. Two studies, one encompassing the densely instrumented region of northern Wisconsin noted above (Desai et al. 2007) and another spanning the deciduous forests of the eastern United States (Albani et al. 2006), ran a terrestrial carbon cycle model that includes successional processes (Moorcroft et al. 2001) over century time scales, and compared the spatially distributed model output to regional observations including flux towers and forest inventory measurements. The time span of both studies allowed the authors to discuss the accuracy of the modeled response of the forest carbon cycle to changes in climate and atmospheric composition, thus working toward evaluating predictive skill. Albani et al. (2006) added a comparison of model predictions to manipulative experiments that investigated ecosystem response to future atmospheric CO_2 concentrations. A notable European study merged multiple flux towers, satellite remote sensing and ecosystem carbon cycle modeling in a study of the 2003 European heat wave

(Reichstein et al. 2006). Studies such as these are converging toward the fully integrated flux upscaling discussed in this chapter, and are applying these regional flux estimates to both diagnosis and prediction.

16.3.3 Flux Upscaling Comparison to and Integration with Inventory Approaches

Integration of landscape-scale forest carbon cycle measurements with flux tower and remote sensing observations into regional to continental flux diagnostics – flux tower upscaling - is a means toward the end of a diagnostic and predictive understanding of the terrestrial carbon cycle. This end will be aided by comparison of flux tower upscaling to, and integration of flux tower upscaling with, the complementary methods of atmospheric and biomass inventories (Fig. 16.1). Examples of such integration are already emerging in the literature, and while these studies do not yet show the full array of measurements envisioned in the flux upscaling described in this chapter, they represent necessary and valuable progress since we do not know precisely what observational array is needed to achieve the goals of the NACP science plan. Janssens et al. (2003) has already been discussed. Global-scale comparisons of decadal-scale atmospheric inventory measurements and land-surface remote sensing foreshadow the potential benefit of integrating these approaches, merging the process understanding involved in ecosystem models driven by climate and remote sensing data with the integral constraint on the large-scale carbon cycle provided by atmospheric inversions (Myneni et al. 1997, Nemani et al. 2003). Formal integration of multiple, long-term measures of the global carbon cycle, such as ocean and atmospheric inventory data (Jacobson et al. 2007a, b) and global temperature, CO_2, land use history and ocean data (Kheshgi and Jain 2003, Ricciuto et al. 2008) do not yet include flux upscaling results. Several efforts are underway to merge atmospheric inversions with flux upscaling methods (e.g. Matross et al. 2006). A major regional experiment, the NACP midcontinental intensive regional study (MCI, Ogle et al. 2006), is synthesizing the data needed for perhaps the most comprehensive merger of atmospheric inventory, biomass inventory and flux upscaling to date.

16.4 Challenges and Needs

The concept of flux upscaling has been with the research community for at least two decades (e.g. Sellers et al. 1992). The complexity of the flux upscaling problem is rooted primarily in the gaps in spatiotemporal coverage of the observational approaches (Fig. 16.1) and the multiple processes contributing to forest-atmosphere carbon exchange (Fig. 16.2). This research problem requires aggressive and ongoing experimentation to enable techniques to mature and become more suited to integration across methods and scales. The community must also keep in mind that the methods envisioned in the NACP implementation plan (Denning et al. 2005) may or may not be precisely those needed to achieve the goals of the NACP science plan (Wofsy and Harriss 2002), and will be revised as we learn how best to achieve the goals of diagnosis, attribution and prediction. We must also keep in mind that while human fossil fuel use is the dominant source of uncertainty in predicting the future greenhouse gas burden of the atmosphere, uncertainty in the future terrestrial carbon cycle is large (Friedlingstein et al. 2006), and improved capacity to diagnose and predict these fluxes will greatly advance our ability to design and implement effective greenhouse gas management strategies. Within these broader requirements, more specific needs have emerged and are summarized below.

16.4.1 Methodological Advances in the Components of the Flux Upscaling Observational System

We can benefit a great deal from creative ways to fill the gaps in methodology illustrated in Figs. 16.1 and 16.2 and observe more completely the processes governing the carbon cycle. Progress may come via new observational methods or new mechanistic understanding that reduces the degrees of freedom in the system. For example, widely accessible remotely sensed data is limited to land cover, land cover change and observations of leaf area and absorbed radiation from which GPP can be inferred. Lidar shows promise for remote observations of NPP (Lefsky et al. 2005) and multi-spectral remote sensing may enable leaf nitrogen to be mapped in addition to leaf area (Asner and Vitousek 2005). Eddy covariance measurements only observe NEE directly, but complementary measurements of tracers such as $^{13}CO_2$ (Bowling et al. 2003) or carbonyl sulfide (COS, Montzka et al. 2007) may enable photosynthetic and respiratory fluxes to be observed independently.

Integration of multiple measurements elevates the need for careful evaluation of measurement uncertainty which, if neglected, will lead to false interpretation of the observations. Examples include questions that remain concerning the offset between eddy covariance and chamber flux measurements (e.g. LaVigne et al. 1997), the lack of energy balance closure in flux tower measurements (e.g. Wilson et al. 2002), and temporal lags between inventory and flux measurements (e.g. Barford et al. 2001). These problems have proven difficult to solve, but progress continues to be made in methodology (Davidson et al. 2002, Staebler and Fitzjarrald 2004, Van Gorsel et al. 2007), quantification of measurement uncertainty (Richardson et al. 2006) and assessment of methodological consistency (Oncley et al. 2007, Gough et al. 2008). Continued evaluation, reduction and accurate documentation of the dominant uncertainties in our observational methods will strengthen our efforts.

The precise composition of the landscape-scale forest carbon cycle sampling tier needs to be considered. For example, while the instrumentation for many of the observations used in this tier is relatively inexpensive, the measurements are often labor intensive and therefore difficult to maintain over time at landscape scales. Eddy-covariance flux measurements, while more expensive initially, can be relatively inexpensive to maintain once established. It may be that relatively lightly-instrumented eddy covariance towers should be considered as part of the landscape-scale measurement tier, and used to complement heavily-instrumented core sites. These clusters of eddy-covariance flux measurement sites can then be enhanced by a spatially extensive sampling of carbon stocks and fluxes described in this manual for short periods of time with the goal of establishing a sound link with remote sensing. The more labor intensive measurements might then be limited to co-location with flux tower sites, with a focus on those needed to identify the mechanisms responsible for the changes observed over time at the flux towers. The landscape-scale forest carbon cycle measurements must be closely coordinated with the intensive measurement tier (core flux tower sites) of the NACP science implementation strategy (Denning et al. 2005). Research that addresses the observational design required for effective flux tower upscaling is essential.

16.4.2 Integration of Flux Upscaling with Complementary Approaches

Since some methodological weaknesses in all of our regional to continental-scale flux estimate approaches exist (Figs. 16.1 and 16.2), we must take advantage of the strengths of the available methods and address the weaknesses with alternative approaches. For example, net ecosystem productivity (NEP) of carbon is strongly linked to land use history, which is highly heterogeneous in space and so is difficult to extrapolate across the landscape. Measurements of NEP may also suffer from systematic uncertainties (lack of below-ground carbon accumulation data in biomass inventories, loss of eddy covariance data in stable situations, limited spatial and temporal resolution for atmospheric inventories). Thus, flux tower upscaling or inventory approaches may individually yield biased or inaccurate measurements of continental-scale NEP. A complementary approach may provide more robust findings. Eddy covariance measurements of NEP, for example, are very precise, so flux towers are well suited to measuring variability in NEP over time. In addition, temporal variability in NEP may be dominated by climate and therefore be relatively coherent over space. Therefore, flux tower upscaling may prove to be an excellent approach to diagnosing temporal variability in regional to continental NEP (e.g. Reichstein et al. 2006). Atmospheric inventories, on the other hand, while difficult to interpret locally in space because of vigorous atmospheric mixing, are extremely accurate at global scales because the entire atmospheric burden of CO_2 can be measured very accurately. Biomass inventories, while difficult to apply globally and lacking in temporal resolution, are effective for regional, decadal scale measures of carbon balance (e.g. CCSP 2007). These approaches are complementary and help us obtain more complete quantification of the terrestrial carbon cycle.

Continuing research that integrates flux upscaling with complementary evaluations of the large-scale carbon cycle is needed to test the strengths and weaknesses of various combinations of observations and obtain a more comprehensive understanding of the terrestrial carbon cycle. Approaches include integration of flux upscaling methods with: (1) measurements from atmospheric and/or biomass inventories, (2) constraints on oceanic and anthropogenic processes, (3) observations of the past time evolution of the climate system and (4) manipulative experiments that explore responses of terrestrial ecosystems to future climate and environmental conditions.

Additional landscape-scale measurements not specific to the carbon cycle may advance the objectives of this measurement tier of the NACP and should be considered. Observations that enable changes in the carbon cycle to be attributed to relevant processes will aid the goals of attribution of the factors governing terrestrial carbon fluxes. Examples include measurements of the hydrologic cycle (e.g. sap flux) and nutrient cycles (e.g. nitrogen). Observations that integrate over space (e.g. stream flow, water table depth) may be especially helpful in flux upscaling, while those that integrate over time (e.g. tree rings) may be particularly useful in achieving the goal of predicting regional fluxes.

16.4.3 Advances in Integration of Landscape-Scale Forest Carbon Cycle Measurements with Flux Towers and Remote Sensing

Integration of the multi-faceted observational system proposed for the flux upscaling component of the NACP implementation plan requires ready access to a high-quality, uniform, well-documented database. Reliable data and metadata reporting by investigators is essential. Proactive data management (e.g. Cook and Thornton 2005) greatly facilitates data integration and synthesis.

Synthesis of data across sites is greatly facilitated by coherent and continuous measurement networks. The majority of NACP flux measurements and landscape-scale forest carbon cycle measurements (Tiers 1 and 2, respectively) are supported by individual, short-term (3–5 year) research projects, sometimes with the expectation that these individual efforts will lead to an integrated whole that spans decadal time scales. This presents a major challenge regarding the integration of these data for flux upscaling over climatological time scales. Individual investigators are rightfully encouraged to experiment with methods, yet this experimentation often results in heterogeneity in observational methods or measurements across sites, which slows integration toward flux upscaling. A compromise between shorter-term experimentation and longer-term, tightly networked measurements needs to be made; the former advances our methods and understanding, while the latter strengthens our ability to synthesize multiple sites and data types across space and time.

Continued research concerning the methods of integration is clearly required since no accepted means of creating a regionally upscaled flux estimate built on flux tower, remote sensing, and landscape-scale forest carbon cycle data yet exists. Regional intensive studies such as the MCI should help to establish these methods. For methodology to be established, it must include internal consistency checks as well as validation by comparison to independent methods. Both types of research are included, for example, in the MCI science plan (Ogle et al. 2006).

Finally, network design research is needed to provide quantitative evaluation of the measurements required for successful flux upscaling efforts and for successful integration of flux upscaling work with atmospheric and biomass inventory methods. Data assimilation methodology has the potential to merge our best understanding of terrestrial carbon cycle processes (including multiple hypotheses concerning how these processes function) and observational methods (including characterization of the uncertainty in these observations) to produce quantitative recommendations concerning observational networks. The work by Hargrove et al. (2003) cited by Hollinger in the opening chapter of this manual sets an important precedent, but only begins the process of evaluating the observational design proposed in the NACP science implementation strategy (Denning et al. 2005). It is also important to keep in mind that the results of such formal network design studies are only as useful as the quality of the information that is input (e.g. model process accuracy, realistic evaluation of observational uncertainties). Field scientists, terrestrial modelers and statisticians must work together closely to advance network design effectively and efficiently.

16.5 Conclusions

The NACP science implementation plan speaks of four tiers of measurements: remote sensing (tier 4), biomass inventory measurements (tier 3), landscape-scale carbon cycle measurements (tier 2) and intensive measurement sites/flux towers (tier 1). Integration of these measurements into regional to continental assessments of terrestrial carbon cycle fluxes has, in the small number of existing published studies, fallen along the lines of biomass inventory studies, flux upscaling efforts that attempt to extrapolate tier 1 measurements across space and atmospheric inventories (inversions). The flux upscaling component of the NACP implementation plan, of which the landscape-scale forest carbon cycle measurements discussed in this manual are a part, arguably presents the most complex integration task of the three approaches to obtaining large scale flux estimates. This complexity emerges from the limits of the observational methods and the number of processes involved in the terrestrial carbon cycle. This component of the NACP, however, provides detailed insight into ecosystem processes and is well suited toward developing a stronger mechanistic understanding of the processes governing the regional carbon cycle. In this way, flux upscaling greatly augments inventory methods that are often more limited in spatial and/or temporal resolution and process understanding.

The complexity of flux upscaling via integration of flux tower data, remote sensing and landscape-scale forest carbon cycle measurements has prevented an established set of methods from emerging to date. Integration of these observations into regional flux estimates is an active topic of research. As a result, the observations deemed essential to this endeavor, and the landscape-scale measurement tier in particular, are likely to evolve as research results emerge. Vigorous interaction among the scientific communities advancing the NACP and careful attention to the overall goals of the NACP (diagnosis, attribution and prediction) is needed to facilitate effective maturation of this research. Network design, data assimilation methods, evaluation of process understanding and advancement of observational methods all have important roles to play in this effort. Development of coherent, long-term networks of observations and readily accessible data and metadata are critical to the success of flux upscaling efforts across time and space.

Acknowledgments This chapter represents the integration of the ideas and efforts of numerous colleagues and students, as well as the intersection of a number of complementary research grants that have allowed my research group to investigate various aspects of flux tower upscaling and carbon cycle science. Sources of support include the U.S. Department of Energy's Office of Science (BER) through the Terrestrial Carbon Processes program, and the Northeastern and Midwestern Regional Centers of the National Institute for Climatic Change Research, the U.S. Department of Commerce, National Oceanic and Atmospheric Administration (NOAA), Office of Global Programs, Global Carbon Cycle program, grant number NA040AR4310124, the Terrestrial Ecology Program of the National Aeronautics and Space Administration, and the National Science Foundation, Division of Environmental Biology. Many thanks are due for the persistence, patience and endless good cheer of the lead editor for this manual, Coeli Hoover.

Literature Cited

Albani M, Medvigy D, Hurtt GC, Moorcroft PR (2006) The contributions of land-use change, CO_2 fertilization, and climate variability to the eastern US carbon sink. Global Change Biology 12: 2370–2390, doi: 10.1111/j.1365-2486.2006.01254

Asner GP, Vitousek PM (2005) Remote analysis of biological invasion and biogeochemical change. Proceedings National Academy of Science 102:4383–4386

Barford CC, Wofsy SC, Goulden ML, Munger JW, Pyle E Hammond, Urbanski SP, Hutyra L, Saleska SR, Fitzjarrald D, Moore K (2001) Factors controlling long- and short-term sequestration of atmospheric CO_2 in a mid-latitude forest. Science 294(5547), 1688–1691

Bond-Lamberty B, Wang C, Gower ST (2004) Net primary production and net ecosystem production of a boreal black spruce wildfire chronosequence. Global Change Biology 10: 473–487, doi: 10.1111/j.1529-8817.2003.0742.x

Bowling DR, Pataki DE, Ehleringer JR (2003) Critical evaluation of micrometeorological methods for measuring ecosystem-atmosphere isotopic exchange of CO_2. Agricultural and Forest Meteorology 116(3–4):159–179

Braswell BH, Sacks WJ, Linder E, Schimel DS (2005) Estimating diurnal to annual ecosystem parameters by synthesis of a carbon flux model with eddy covariance net ecosystem exchange observations. Global Change Biology 11(2):335–355

CCSP (2007) The First State of the Carbon Cycle Report (SOCCR): North American Carbon Budget and Implications for the Global Carbon Cycle. A report by the U.S. Climate Change Science Program and the Subcommittee on Global Change Research. King AW, Dilling L,

Zimmerman GP, Fairman DM, Houghton RA, Marland GH, Rose AZ, Wilbanks TJ (eds.) National Ocean and Atmospheric Administration, National Climatic Data Center, Asheville, NC, 242 p

Cohen WB, Maiersperger TK, Turner DP, Ritts WD, Pflugmacher D, Kennedy RE, Kirschbaum A, Running SW, Costa M, Gower ST (2006) MODIS land cover and LAI collection 4 product quality across nine sites in the western hemisphere. IEEE Transactions in Geosciences and Remote Sensing 44: 1843–1857

Cook RB, Thornton, P (2005) Data management for the North American Carbon Program. Workshop Report. New Orleans, Louisiana. Jan. 25–27, 2005, 48 p (http://www.nacarbon.org).

Davidson EA, Savage K, Verchot LV, Navarro R (2002) Minimizing artifacts and biases in chamber-based measurements of soil respiration. Agricultural and Forest Meteorology 113: 21–37

Denning AS, Oren R, McGuire D, Sabine C, Doney S, Paustian K, Torn M, Dilling L, Heath L, Tans P, Wofsy S, Cook R, Waltman S, Andrews A, Asner G, Baker J, Bakwin P, Birdsey R, Crisp D, Davis K, Field C, Gerbig C, Hollinger D, Jacob D, Law B, Lin J, Margolis H, Marland G, Mayeux H, McClain C, McKee B, Miller C, Pawson S, Randerson J, Reilly J, Running S, Saleska S, Stallard R, Sundquist E, Ustin S, Verma S (2005) Science Implementation Strategy for the North American Carbon Program. Prepared for the U.S. Carbon Cycle Science Steering Group and the Interagency Working Group On Carbon (http://www.nacarbon.org)

Desai AR, Noormets A, Bolstad PV, Chen J, Cook BD, Davis KJ, Euskirchen ES, Gough C, Martin JM, Ricciuto DM, Schmid HP, Tang J, Wang W (2008) Influence of vegetation and seasonal forcing on carbon dioxide fluxes across the Upper Midwest, USA: Implications for regional scaling. Agricultural and Forest Meteorology 148: 288–308, doi:10.1016/j.agrformet.2007.08.001

Desai AR, Moorcroft PR, Bolstad PV, Davis KJ (2007) Regional carbon fluxes from an observationally constrained dynamic ecosystem model: Impacts of disturbance, CO_2 fertilization, and heterogeneous land cover. Journal of Geophysical Research 112:G01017, doi:10.1029/2006JG000264.

Friedlingstein P, Cox PM, Betts RA, Bopp L, von Bloh W, Brovkin V, Cadule P, Doney SC, Eby M, Fung IY, Bala G, John J, Jones CD, Joos F, Kato T, Kawamiya M, Knorr W, Lindsay K, Matthews HD, Raddatz T, Rayner P, Reick C, Roeckner E, Schnitzler K-G, Schnur R, Strassmann K, Weaver AJ, Yoshikawa C, Zeng N (2006) Climate-carbon cycle feedback analysis: Results from the C^4MIP model intercomparison. Journal of Climate 19: 3337–3353

Gough CM, Vogel CS, Schmid HP, Su H-B, Curtis PS (2008) Multi-year convergence of biometric and meteorological estimates of forest carbon storage. Agricultural and Forest Meteorology 148: 158–170, doi:10.1016/j.agrformet.2007.08.004

Goulden ML, Winston GC, McMillan AMS, Litvak ME, Read EL, Rocha AV, Elliot JR (2006) An eddy covariance mesonet to measure the effect of forest age on land–atmosphere exchange. *Global Change Biology* 12: 2146–2162, doi: 10.1111/j.1365-2486.2006.01251.x

Goulden M, Munger JW, Fan SM, Daube BC, Wofsy SC (1996) Measurements of carbon sequestration by long-term eddy covariance: Methods and a critical evaluation of accuracy. Global Change Biology 2(3):169–182

Gurney KR, Law RM, Denning AS, Rayner PJ, Baker D, Bousquet P, Bruhwiler L, Chen, YH, Ciais P, Fan S, Fung IY, Gloor M, Heimann M, Higuchi K, John J, Maki T, Maksyutov S, Masarie K, Peylin P, Prather M, Pak BC, Randerson J, Sarmiento J, Taguchi S, Takahashi T, Yuen CW (2002) Towards robust regional estimates of CO_2 sources and sinks using atmospheric transport models. Nature 415: 626–630

Hanson PJ, Amthor JS, Wullschleger SD, Wilson KB, Grant RF, Hartley A, Hui D, Hunt ER Jr, Johnson DW, Kimball JS, King AW, Luo Y, McNulty SG, Sun G, Thornton PE, Wang S, Williams M, Baldocchi DD, Cushman RM (2004) Oak forest carbon and water simulations: Model intercomparisons and evaluations against independent data. Ecological Monographs 74(3):443–489

Hargrove WW, Hoffman FM, Law BE (2003) New analysis reveals representativeness of the AmeriFlux network. EOS Transactions AGU 84(48):529

Heinsch FA, Zhao M, Running SW, Kimball JS, Nemani RR, Davis KJ, Bolstad PV, Cook BD, Desai AR, Ricciuto DM, Law BE, Oechel WC, Kwon H, Luo H, Wofsy SC, Dunn AL, Munger JW, Baldocchi DD, Xu L, Hollinger DY, Richardson AD, Stoy PC, Siqueira MBS, Monson RK, Burns S, Flanagan LB (2006) Evaluation of remote sensing based terrestrial productivity from MODIS using regional tower eddy flux network observations. IEEE Transactions on Geosciences and Remote Sensing 44: 1908–1925

Horst TW, Weil JC (1992) Footprint estimation for scalar flux measurements in the atmospheric surface layer. Boundary Layer Meteorology 59: 279–296

Houghton RA (2003) Revised estimates of the annual net flux of carbon to the atmosphere from changes in land use and land management 1850–2000. Tellus Series B-Chemical and Physical Meteorology 55(2):378–390

Jacobson AR, Mikaloff Fletcher SE, Gruber N, Sarmiento JL, Gloor M (2007a) A joint atmosphere-ocean inversion for surface fluxes of carbon dioxide: 1. Methods and global-scale fluxes. Global Biogeochemical Cycles 21:GB1019, doi: 10.1029/2005GB002556

Jacobson AR, Mikaloff Fletcher SE, Gruber N, Sarmiento JL, Gloor M (2007b) A joint atmosphere-ocean inversion for surface fluxes of carbon dioxide: 2. Regional results. Global Biogeochemical Cycles 21:GB1020, doi:10.1029/2006GB002703

Jain AK, Yang X (2005) Modeling the effects of two different land cover change data sets on the carbon stocks of plants and soils in concert with CO_2 and climate change. Global Biogeochemical Cycles 19: GB2015, doi:10.1029/2004GB002349

Janssens IA, Freibauer A, Ciais P, Smith P, Nabuurs, G-J, Folberth G, Schlamadinger B, Hutjes RWA, Ceulemans R, Schulze E-D, Valentini R, Dolman AJ (2003) Europe's terrestrial biosphere absorbs 7 to 12% of European anthropogenic CO_2 emissions. Science 300: 1538–1542

Kheshgi HS, Jain AK (2003) Projecting future climate change: Implications of carbon cycle model intercomparisons. Global Biogeochemical Cycles 17(2):1047, doi: 10.1029/2001GB001842

Knorr W, Kattge J (2005) Inversion of terrestrial ecosystem model parameter values against eddy covariance measurements by Monte Carlo sampling. Global Change Biology 11: 1333–1351, doi: 10.1111/j.1365-2486.2005.00977.x

Kucharik CJ, Barford CC, Maayar M El, Wofsy SC, Monson RK, Baldocchi DD (2006) A multi-year evaluation of a dynamic global vegetation model at three AmeriFlux forest sites: Vegetation structure, phenology, soil temperature, and CO_2 and H_2O vapor exchange. Ecological Modelling 196(1–2):1–31

LaVigne MB, Ryan MG, Anderson DE, Baldocchi DD, Crill PM, Fitzjarrald DR, Goulden ML, Gower ST, Massheder JM, McCaughey JH, Rayment M, Striegl RG (1997) Comparing nocturnal eddy covariance measurements to estimates of ecosystem respiration made by scaling chamber measurements at six coniferous boreal sites. Journal of Geophysical Research - Atmospheres 102(D24):28977–28985

Lefsky MA, Turner DP, Guzy M, Cohen WB (2005) Combining lidar estimates of aboveground biomass and Landsat estimates of stand age for spatially extensive validation of modeled forest productivity. Remote Sensing of Environment 95(4):549–558, doi: 10.1016/j.rse.2004.12.022.

Loescher HW, Law BE, Mahrt L, Hollinger DY, Campbell J, Wofsy SC (2006) Uncertainties in, and interpretation of, carbon flux estimates using the eddy covariance technique. Journal of Geophysical Research 111:D21S90, doi: 10.1029/2005JD006932

Matross DM, Andrews A, Mahadevan P, Gerbig C, Lin JC, Wofsy SC, Daube BC, Gottlieb EW, Chow VY, Lee JT, Zhao C, Bakwin PS, Munger JW, Hollinger DY (2006) Estimating regional carbon exchange in New England and Quebec by combining atmospheric, ground-based and satellite data. Tellus B 58(5):344–358, doi: 10.1111/j.1600-0889.2006.00206.x

Montzka SA, Calvert P, Hall BD, Elkins JW, Conway TJ, Tans PP, Sweeney C (2007) On the global distribution, seasonality, and budget of atmospheric carbonyl sulfide (COS) and some similarities to CO_2. Journal of Geophysical Research 112:D09302, doi: 10.1029/2006JD007665

Moorcroft PR, Hurtt GC, Pacala SW (2001) A method for scaling vegetation dynamics: The ecosystem demography model (ED). Ecological Monographs 71: 557–585

Myneni RB, Keeling CD, Tucker CJ, Asrar G, Nemani RR (1997) Increased plant growth in the northern high latitudes from 1981 to 1991. Nature 386: 698–702

Nemani RR, Keeling CD, Hashimoto H, Jolly WM, Piper SC, Tucker CJ, Myneni RB, Running SW (2003) Climate-driven increases in global terrestrial net primary production from 1982 to 1999. Science 300: 1560–1563

Ogle S, Davis K, Andrews A, Gurney K, West T, Cook R, Parkin T, Morisette J, Verma S, Wofsy S (2006) Science Plan: Mid-Continent Intensive Campaign of the North American Carbon Program. Prepared for the carbon cycle Science Steering Group and the Interagency Working Group on Carbon (http://www.nacarbon.org)

Oncley SP, Foken T, Vogt R, Kohsiek W, DeBruin HAR, Bernhofer C, Christen A, van Gorsel E, Grantz D, Feigenwinter C, Lehner I, Liebethal C, Liu H, Mauder M, Pitacco A, Ribeiro L, Weidinger T (2007) The Energy Balance Experiment EBEX-2000. Part I: Overview and energy balance. Boundary-Layer Meteorology 123: 1–28, doi: 10.1007/s10546-007-9161-1

Pacala SW, Hurtt GC, Baker D, Peylin P, Houghton RA, Birdsey RA, Heath LS, Sundquist E, Stallard RF, Ciai P, Moorcroft P, Caspersen JP, Shevliakova E, Moore B, Kohlmaier G, Holland EA, Gloor M, Harmon ME, Fan S-M, Sarmiento JL, Goodale CL, Schimel D, Field CB (2001) Consistent land- and atmosphere-based U.S. carbon sink estimates. Science 292: 2316–2322

Reichstein M, Ciais P, Papale D, Valentini R, Running S, Viovy N, Cramer W, Granier A, Ogee J, Allard V, Aubinet M, Bernhofer C, Buchmann N, Carrara A, Grunwald T, Heimann M, Heinesch B, Knohl A, Kutsch W, Loustau D, Manca G, Matteucci G, Miglietta F, Ourcival JM, Pilegaard K, Pumpanen J, Rambal S, Schaphoff S, Seufert G, Soussana JF, Sanz MJ, Vesala T, Zhao M (2006) Reduction of ecosystem productivity and respiration during the European summer 2003 climate anomaly: a joint flux tower, remote sensing and modelling analysis. Global Change Biology 12: 1–18

Ricciuto DM, KJ Davis, K Keller (2008) A Bayesian synthesis inversion of carbon cycle observations: How can observations reduce uncertainties about future sinks? Global Biogeochemical Cycles (in press), doi: 10.1029/2006GB002908

Richardson AD, Hollinger DY, Burba GG, Davis KJ, Flanagan LB, Katul GG, Munger JW, Ricciuto DM, Stoy PC, Suyker AE, Verma SB, Wofsy SC (2006) A multi-site analysis of random error in tower-based measurements of carbon and energy fluxes. Agricultural and Forest Meteorology 136: 1–18

Staebler RM, Fitzjarrald DR (2004) Observing subcanopy CO_2 advection. Agricultural and Forest Meteorology 122: 139–156

Sellers PJ, Hall FG, Asrar G, Strebel DE, Murphy RE (1992) An overview of the First International Satellite Land Surface Climatology Project (ISLSCP) Field Experiment (FIFE). Journal of Geophysical Research 97: 18345–18371

Tang J, Bolstad PV, Desai AR, Martin JM, Cook BD, Davis KJ, Carey EV (2008) Ecosystem respiration and its components in an old-growth northern forest. Agricultural and Forest Meteorology 148: 171–185, doi: 10.1016/j.agrformet.2007.08.008.

Thornton PE, Law BE, Gholz HL, Clark KL, Falge E, Ellsworth DS, Goldstein AH, Monson RK, Hollinger D, Falk M, Chen J, Sparks JP (2002) Modeling and measuring the effects of disturbance history and climate on carbon and water budgets in evergreen needleleaf forests. Agricultural and Forest Meteorology 113: 185–222

Turner DP, Ritts WD, Law BE, Cohen WB, Yang Z, Hudiburg T, Campbell JL, Duane M (2007) Scaling net ecosystem production and net biome production over a heterogeneous region in the western United States. Biogeosciences Discussions 4: 1–43

Turner DP, Ritts WD, Cohen WB, Gower ST, Running SW, Zhao M, Costa MH, Kirschbaum A, Ham J, Saleska S, Ahl DE (2006) Evaluation of MODIS NPP and GPP products across multiple biomes. Remote Sensing of Environment 102: 282–292

Van Gorsel E, Leuning R, Cleugh HA, Keith H, Suni T (2007) Nocturnal carbon efflux: reconciliation of eddy covariance and chamber measurements using an alternative to the u*-threshold filtering technique. Tellus 59B:397–403

Wang W, Davis KJ, Cook BD, Butler MP, Ricciuto DM (2006) Decomposing CO_2 fluxes measured over a mixed ecosystem at a tall tower and extending to a region: A case study. Journal of Geophysical Research Biogeosciences 111(G2):G02005, doi: 10.1029/2005JG000093

Wilson K, Goldstein A, Falge E, Aubinet M, Baldocchi D, Berbigier P, Bernhofer C, Ceulemans R, Dolman H, Field C, Grelle A, Ibrom A, Law BE, Kowalski A, Meyers T, Moncrieff J, Monson R, Oechel W, Tenhunen J, Valentini R, Verma S (2002) Energy balance closure at FLUXNET sites. Agricultural and Forest Meteorology 113: 223–243

Wofsy SC, Harriss RC (2002) The North American Carbon Program (NACP). Report of the NACP Committee of the U.S. Interagency Carbon Cycle Science Program. Washington, DC: US Global Change Research Program (http://www.nacarbon.org)

Chapter 17
Landscape-Scale Carbon Sampling Strategy – Lessons Learned

John B. Bradford, Peter Weishampel, Marie-Louise Smith, Randall Kolka, David Y. Hollinger, Richard. A. Birdsey, Scott Ollinger, and Michael G. Ryan

Abstract Previous chapters examined individual processes relevant to forest carbon cycling, and characterized measurement approaches for understanding those processes at landscape scales. In this final chapter, we address our overall approach to understanding forest carbon dynamics over large areas. Our objective

J.B. Bradford
US Forest Service, Northern Research Station, 1831 Hwy 169 E., Grand Rapids, MN 55744
E-mail: jbbradford@fs.fed.us

P. Weishampel
University of Minnesota, Department of Soil, Water, and Climate, 1991 Upper Buford Circle, St. Paul, MN 55108
E-mail: peter.weishampel@gmail.com

M.-L. Smith
US Forest Service, Northern Research Station, Current: US Forest Service, Legislative Affairs, 201 14th Street, SW Washington, DC 20250-1130
E-mail: marielouisesmith@fs.fed.us

R. Kolka
US Forest Service, Northern Research Station, 1831 Hwy 169 E, Grand Rapids, MN 55744
E-mail: rkolka@fs.fed.us

D.Y. Hollinger
US Forest Service, Northern Research Station, 271 Mast Road, Durham, NH 03824
E-mail: dhollinger@fs.fed.us

R.A. Birdsey
US Forest Service, Northern Research Station, 11 Campus Blvd., Suite 200, Newtown Square, PA 19073
E-mail: rbirdsey@fs.fed.us

S. Ollinger
Complex Systems Research Center, University of New Hampshire, 56 College Road, Durham, New Hampshire 03824-3589
E-mail: scott.ollinger@unh.edu

M.G. Ryan
US Forest Service, Rocky Mountain Research Station, 240 W. Prospect Ave., Fort Collins, CO 80526
E-mail: mgryan@fs.fed.us

C.M. Hoover (ed.) *Field Measurements for Forest Carbon Monitoring*,

is to identify any lessons that we learned in the course of measuring a wide range of carbon-related processes in a suite of forested sites. We focus on characterizing the costs and benefits of measuring individual processes and we examine the advantages and limitations to our plot layout. In addition, we draw upon the experience at individual sites to identify important lessons that may be specific to particular forest types or regions.

Keywords Terrestrial carbon cycling, carbon storage, net ecosystem carbon balance, spatial and temporal scaling

17.1 Introduction

Our objective in this chapter is to communicate some initial lessons about the practical challenges of designing and conducting landscape-scale measurements of carbon pools and fluxes. We stress that these conclusions are preliminary; much of our data is still being collected and analyzed and many of the lessons that we will learn from this project are only beginning to become clear. Nevertheless, our experiences provide insight into potential improvements for similar future efforts. Specifically, we address two topics: the cost and benefit of various measurements and the advantages and disadvantages of our plot layout strategy.

17.2 Measurement Costs and Benefits: "What We Measured"

Quantifying carbon pools and fluxes at landscape scales requires identifying the ecological processes that play an important role in the carbon cycle and developing a feasible approach to measure those processes. Individual process measurements can be evaluated by both their contribution to accurate estimates of whole ecosystem carbon dynamics versus their cost in terms of time and money. To critically examine the measurements that we conducted, we present an objective approach to assessing cost and benefit, illustrate how the measurements that we conducted fit qualitatively into that framework for one of our sites, and conclude by considering how measurements can also be valuable for comparison with alternative approaches to carbon accounting.

17.2.1 Costs and Benefits Defined

Developing a sampling strategy for assessing landscape-scale carbon dynamics requires objectively assessing the costs and benefits of various potential measurements. One objective approach to assessing the benefit of measuring a process is to quantify how much the process influences either total ecosystem carbon storage or net ecosystem carbon balance. For the goal of quantifying total carbon stocks, the benefit of measuring individual pools is roughly proportional to the size of the pool. Pools with very small amounts of carbon will contribute very little to overall estimates of carbon storage, and errors in those pools will have only minor consequences. Similarly, for estimating net ecosystem carbon balance, the benefit of an individual carbon flux is equivalent to the size of the flux. Larger carbon fluxes have greater overall influence on total carbon balance and can thus be considered more beneficial to an overall carbon assessment strategy. Relating the benefit of a given measurement to the magnitude of the carbon pool or flux that it measures provides an objective quantification of importance that can be balanced against the cost of the measurement.

In general, the cost of measuring a process can be inferred from the amount of variability that the process displays both across space and time. Although some measurements are inherently more difficult than others, we found that, for long-term measurements at the landscape scale, sampling requirements for characterizing variability overshadowed differences in measurement difficulty. If a process is relatively consistent across the landscape, then only a few plots are sufficient to develop an accurate landscape-scale estimate. Likewise, if the process is consistent through time, infrequent measurements are adequate to characterize temporal patterns. Highly variable processes, on the other hand, require either many plots and/or frequent measurement to characterize spatial or temporal patterns, respectively. We found that measurements requiring multiple visits per year required substantial labor and money. For example, soil CO_2 and methane efflux and litterfall required that we visit plots multiple times per year. By contrast, the very infrequent measurements were time consuming initially, but once completed incurred essentially no additional cost. Live tree and coarse wood biomass, for example, change relatively slowly and consequently required only a single visit to characterize. Recognizing the high cost of repeated measurements is especially relevant considering the short-term nature of most research grants in which intensive work may be possible for a relatively brief period, followed by a longer period of minimal resources. In some cases, measurement of long-term processes can be designed to fit within this funding reality. For example, measuring tree growth can be achieved with either very infrequent (once in several years) measurements combined with increment coring, or with annual diameter measurements. On the other hand, if long-term monitoring resources are available, repeating detailed tree measurements may have advantages. By defining cost and benefit in terms of variability and magnitude, this conceptual framework provides an objective mechanism for evaluating the necessity of measurements, and for beginning to assess sampling design and intensity.

17.2.2 Costs vs. Benefits

Although we are still working to characterize the costs and benefits of measuring various processes, our preliminary results suggest some lessons (Fig. 17.1). Processes

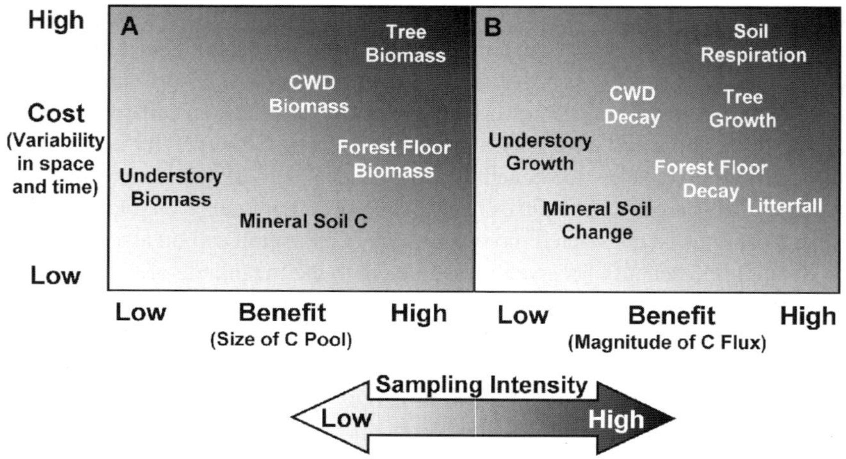

Fig. 17.1 A conceptual framework for characterizing the cost and benefit of measuring individual carbon pools (A) and fluxes (B) in an assessment of landscape-scale carbon dynamics. Benefit is defined as the size of the pool or magnitude of the flux, which is a general measure of the influence a process exerts over ecosystem carbon dynamics. Cost depends primarily on the variability of the process in space and time, with highly variable processes requiring either a large number of plots or high sampling frequency, respectively. This framework provides a mechanism of evaluating the sampling intensity necessary to accurately assess individual pools and fluxes. Processes with greater variability and benefit will require and warrant greater sampling effort, illustrated by the dark background in the upper right of both figures. Placement of individual processes within this framework is shown for subalpine Rocky Mountain forests as an example and will vary between sites. Note that benefits considered here do not include comparability to other approaches like eddy covariance or simulation models

that are low in both influence and variability, notably mineral soil carbon stocks (Chapter 10) at some locations (i.e. Rocky Mountain sites) could be adequately assessed with fewer plots than we utilized. Although mineral soil holds substantial carbon, it is very consistent across space and changes very slowly. Some processes, including understory biomass (Chapter 5) and soil respiration (Chapter 11), incurred high cost with only marginal contribution to our overall assessment of carbon storage or balance. In the forests we examined, understory biomass and productivity was modest, yet required substantial time to quantify. Likewise, soil respiration was one of the most time consuming measurements we initiated (due to instrumentation requirements and necessity of frequent measurement). Although soil respiration was not utilized in the mass balance carbon dynamics approach we adopted, the amount of carbon released via soil respiration is very substantial and provides unique and valuable insight into belowground carbon cycling and storage (Ryan and Law 2005). Tree biomass (Chapter 4) and growth, as major components of carbon cycling in forest systems, were very important for quantifying carbon pools and fluxes, although tree growth was also extremely variable in space and thus costly to measure accurately. We found that the biomass and decomposition of coarse woody debris (Chapters 6 and 9) and forest floor material (Chapter 10) were of intermediate importance and that the cost of effectively measuring coarse woody debris was extremely large due to high spatial variability.

17.3 Comparison with Other Approaches

An important goal of our landscape-scale carbon measurements that is not considered in the above cost-benefit analysis is comparability with other approaches to assessing forest carbon dynamics, notably eddy covariance techniques (Barford et al. 2001, Curtis et al. 2002, Baldocchi 2003) and ecological simulation models that utilize remote sensing data (Turner et al. 2004, Zheng et al. 2006). We found that these comparisons can be hindered by incompatibility in the spatial scale of measurements, the temporal scale of measurements, and the specific processes that are measured. Comparison with eddy covariance data must ensure that the field plots represent the spatial extent of the flux tower footprint, that high frequency eddy covariance measurements can be synthesized to match longer-term field measurement and that the processes quantified by both approaches can be directly compared (see Chapter 16 for details). For example, preliminary comparisons of field measurements with eddy covariance data at Niwot Ridge were limited because the decomposition rate of detrital material is represented by field measurements only as a long-term mean. Consequently, although we found good agreement between NPP estimates from both eddy covariance data and our biometric data (J. Bradford unpublished data), we were unable to directly compare total carbon balance, which is the primary process that is actually measured by eddy covariance techniques (Baldocchi 2003).

Temporal integration with models based on remote sensing data is less complicated because field measurements assess carbon fluxes at the seasonal or annual scale, which is comparable to many simulation models. However, spatial compatibility with the remote sensing data remains a challenge. Although methods have been developed for assessing ground conditions to compare with remote sensing data (e.g. Turner et al. 2005) our plot layout (nested 8–10 m circular plots) was not ideally suited for compatibility with relatively high resolution remote sensing data (i.e. 30 m resolution Landsat), or more coarse resolution data (250 m or 1 km resolution). Comparison with higher resolution data would require characterizing processes over contiguous areas larger than our plots (Grunblatt 1987). Comparison with coarse resolution data requires summarizing our plot data to represent a larger area, which can be accomplished with either a simple mean (assumes the plots are representative of the landscape) or a weighted mean (which relies on outside information about cover type proportions within the landscape). Having an independent classification of the study area would help characterize how well the plots represent the landscape and provide quantification of cover type proportions.

17.4 Plot Layout: "How We Measured it"

For this project, our plots were grouped into FIA-like clusters of four subplots (Bechtold and Patterson 2005) and established in a regular grid across the landscape (Chapter 1). This design is only one of several potential approaches to orienting plots across a landscape. Alternative possibilities include one or more of the following: (a) orienting the plots along one or more transects across the landscape, possibly spanning environmental gradients, vegetation types or patch edges (e.g. Chen et al. 1992); (b) stratifying the landscape into discrete classes *a priori* and establishing plots in each class (Wagner and Fortin 2005); (c) clustering plots at varying spatial scales to facilitate characterization of spatial variability (Rossi et al. 1992); and (d) employing different plot size and/or number of plots for measurement of different processes. To examine the plot layout that we utilized, we present our initial impressions and insights in terms of advantages, limitations, and potential alternative layouts.

17.4.1 Advantages of Our Approach

The two primary advantages of our plot layout design are: (1) the consistency and therefore comparability it facilitates across sites and with FIA data, and (2) the completely unbiased selection of plot locations. Because we established plots very similar to the protocol employed by the FIA program, comparison of our results with FIA results will be very straightforward. Although we are only beginning to explore the possibility of merging our results with FIA data, this integration is likely to prove valuable considering our goal of generating large-area estimates of carbon pools and fluxes. In addition, because our plots are so similar to FIA plots, our work can provide insights about the FIA approach in general, including characterizing the advantages of establishing FIA plots at higher density, quantification of how well individual FIA plots represent the surrounding landscape, and estimates of the most efficient number of subplots to establish per FIA plot. Another advantage of the grid system that we utilized was its consistency across all sites. This consistency enabled direct comparison of spatial variability and patterns between sites that would be substantially more difficult if plot orientation varied between sites. In addition, because we measured all processes at each of our plots (rather than measuring some variables at only a subset of plots) we had a high sample size for comparison between variables.

Perhaps the greatest advantage of our plot layout is the systematic, unbiased process in which plot locations were selected. Whereas most efforts to characterize ecological processes at large scales involve some stratification prior to plot selection (e.g. Hansen et al. 2000), our plot locations were identified simply by a grid overlaid on the landscape. At some sites (notably the 3 Rocky Mountain landscapes), we were frequently surprised by the specific plot locations identified, initially finding some of them to be "unrepresentative" of the landscape as a whole. This suggests that we (and perhaps other researchers) had a notion of how a given forest stand of particular species and age should appear, and locations that differ from that ideal are considered anomalous, even if those anomalies comprise a large proportion of the landscape. This completely objective approach to plot selection provides an opportunity to challenge our biases and potentially identify previously unappreciated landscape conditions that make an important contribution to large-scale carbon dynamics. Although the advantage of comparability to FIA data may not be of interest to all future studies, creating consistency across sites may be increasingly valuable as interest in national-scale carbon accounting grows, and the confrontation of unconscious bias imposed by the systematic plot selection is always useful.

17.4.2 Limitations to Our Approach

Despite these advantages, our initial experiences suggest several limitations to our plot layout. One potential limitation of our approach is that the cluster of four subplots per plot appears to not be the most efficient means of characterizing many of the processes we measured. In a nested sampling scheme like the one we employed, accuracy of the whole-landscape process estimate can be calculated from the within plot variability (variation between subplots within a plot) and the between plot variability (variation between plot means) (Cochran 1977). At all sites, the vast majority of processes were most efficiently measured by establishing independent subplots (rather than subplots clustered into plots with more than 1 subplot), suggesting that our sampling would have been more efficient if we had altered how we located our plots (Bradford et al. in review).

Another important limitation is the potential of plots to be located in areas that are genuinely un-representative of the landscape. The problem can be avoided if a sufficiently large number of plots are established. Although a very large number of plots established on a grid is probably the optimal strategy (ensures representation yet avoids bias), it is practically unfeasible in a reasonably large landscape. Evaluating how well a set of plots represents the landscape requires some type of classification and has not yet been examined for our sites. Another limitation of our gridded plot layout is the potential for plots to be either located very near or far from roads. Plots very near roads are vulnerable to vandalism, which we experienced at multiple sites, and plots extremely far from roads can be difficult to access, substantially increasing the effort required to sample those plots.

17.5 Alternative Plot Layouts

Comparing our plot layout with alternative layouts illustrates four additional limitations to our approach. First, our plots were located on a grid with 250 m between plot centers, meaning that we had no subplot pairs separated by distances between roughly 50 and 200 m and no subplots closer than 35 m. Initial examination of the scale of spatial variability in our measurements suggests that much of the variation occurs at small scales (Bradford et al. in review), and that the 50–200 m range is crucial for geostatistical analysis (D. Hollinger and J. Bradford unpublished data). Consequently, our ability to characterize patterns in spatial variability would have been enhanced by having some subplots separated by a distance of less than 35 m and between 50 and 200 m.

Second, stratifying the landscape prior to plot establishment, a common practice in other landscape-scale carbon assessments (Noormets et al. 2006), may have increased the representation and efficiency of our measurements. Although the results of using an *a priori* stratification are heavily dependent on the assumptions made in the initial stratification, it would have allowed us to partition the landscape into classes like forested versus meadow (at GLEES), peatland versus upland (at Marcell) or young versus old (at Fraser and Bartlett) and we then could have ensured that we have sufficient plots in each class to obtain accurate within-class estimates. This would have minimized the possibility of establishing more plots than are necessary in a particular class and allowed us to efficiently utilize our sampling resources.

Third, our plot layout did not allow us to robustly characterize the importance of edges between vegetation types on carbon dynamics (although some subplots are on or near edges, they were not systematically oriented to allow us to quantify edge effects). Edges are increasingly recognized as important drivers of ecosystem processes, including productivity and carbon cycling (Euskirchen et al. 2006), and by establishing plots along a transect that crosses edges, we could have quantified this importance and potentially strengthened our landscape-scale estimates.

Fourth, we utilized the same number, and roughly same size of plots to characterize most of the processes that we examined. An alternative approach would be to vary the number and/or size of plots, depending on the process variability. For example, we found that in many of our sites, carbon stored in forest floor varied only slightly across the landscape and could have been assessed with fewer plots. Coarse woody debris biomass, by contrast, was both highly variable and non-normally distributed across our landscapes, and consequently may require larger plots and/or more plots within the landscape. By measuring all variables at all plots, we were not as efficient as possible.

17.6 Site Specific Lessons

17.6.1 Bartlett Experimental Forest

Heterogeneity in carbon pools and fluxes at the Bartlett Experimental Forest, a 1,052 ha secondary successional mixed northern hardwood-conifer site in the central White Mountains of New Hampshire, was observed most strongly with respect to gradients in elevation. Though simply expressed, this relationship may not be indicative of a simple pattern. Many factors change with increasing elevation including: decreased air and soil temperatures, increased precipitation, soil type, prevalence of rock fragments and ledges, changes in vegetation composition and thus litter quality and nutrient cycling. From our results to date we cannot conclude which variables play the most important role in the observed elevational trends. We can conclude that the original 1 km^2 sampling frame established in 2004 was inadequate to capture the range in elevation at BEF. However, the aggregated observation of plot-based C flux within the 1 km^2 frame is comparable with that estimated by means of an CO_2 eddy covariance tower located at the center of the 1 km^2 sampling area. The sampling frame at BEF was expanded in 2005 to include 11 additional plots spanning the range in elevation at BEF (210–915 m).

Our observations of individual C pools and fluxes across the BEF are similar to those observed in previous studies in the northeastern region. Relationships among C pools and fluxes as well as to other ecosystem characteristics such as ANPP, aboveground biomass, litter quality, and N cycling, are more variable. Additional plots stratified to cover gaps in elevation, species composition, and disturbance history, as well as a longer temporal sequence of measurements, will allow better assessment of trends into the future.

17.6.2 Marcell Experimental Forest

Our implementation the landscape sampling strategy at the Marcell Experimental Forest (MEF) was hampered by heterogeneity of cover types. The MEF landscape is composed of multiple watersheds that contain upland and peatland portions. Vegetation cover in both upland and peatland portions is highly variable. While uplands are mostly aspen-dominated, there are distinct patches of mixed hardwoods and conifers. Peatlands included forested and non-forested bog and fen communities. Because of the high carbon storage known to exist in peatland soils, we elected to include them in our sampling through non-biased plot and sub-plot establishment. However, the distribution of forested and peatland cover types across the sampling plots did not represent the landscape comprehensively. For example, forested peatlands cover approximately 10% of the land area in the 1-km^2 sampling grid at MEF, but non-biased plot selection yielded only one forested peatland subplot out of the 64 FIA subplots established within the sampling frame. Consequently, in heterogeneous landscapes, a stratified sampling approach that guarantees inclusion of important landscape elements is desirable.

In heterogeneous landscapes, no single approach to sampling C pools and fluxes may be useful for all sample plots. The inclusion of peatlands in our analysis immediately suggested a need for different approaches to sampling C pools and fluxes within subplots, as storage and movement of carbon in peatlands can be quite different from uplands. Peatlands contain a greater depth of organic carbon than is found in upland forest, so we sampled peat deeper than we sampled mineral soil, coring up to 5 m deep, or until contact was made with hard mineral soil. Much of NPP (net primary productivity) in peatlands can occur as *Sphagnum* moss growth, so we attempted to measure these C fluxes on peatland subplots using biometric approaches. Lastly, wetlands are the largest natural source of atmospheric methane. While methane flux represents a small fraction of the ecosystem C balance, we attempted to monitor methane because of its importance as a greenhouse gas.

Each of the additional storage and flux measurements measured on peatland subplots can be analyzed in terms of cost and benefits, as outlined previously and in Fig. 17.1. For example, the C contained in peat is substantial, and is close to 50% of the carbon stored in our 1-km^2 sampling area, despite the fact that peatlands represent only 18% of the subplots. The depth of peat requires a greater sampling effort than mineral soil, but repeated sampling is not required. C storage in peat is highly sensitive to variability in peat depth between subplots. Thus, assessing peat C might be considered a moderate cost, high benefit measurement. *Sphagnum* moss NPP measurements has high costs, as spatial variability imposed by microtopography requires repeated measurements in space and repeated visits. Because *Sphagnum* NPP can rival tree NPP in peatlands, there is a potentially high benefit to these measurements. Lastly, methane fluxes are notoriously variable in space and time but because peatlands are a dominant source of methane in the landscape, their measurement carries a high benefit as well as a high cost.

17.6.3 Subalpine Rocky Mountains

We examined carbon pools and fluxes at three separate small subalpine forested landscapes in the southern Rocky Mountains. Spatial heterogeneity in these forests is driven by elevation, ecosystem type and disturbance history. One of our sites was severely burned >300 years ago and was strip-logged approximately 50 years ago, one was clearcut roughly 100 year ago and contains some aspen stands in addition to the dominant coniferous forest, and one has been unmanaged but consists of a mosaic of forests and alpine meadows. At all sites, species composition varies along an elevational gradient from lower elevation lodgepole pine to higher elevation spruce-fir forests. This variability created dramatic differences in carbon storage and cycling across the landscape in patterns that were not easily predicted. Accounting for this variability was our primary challenge in assessing landscape-scale carbon dynamics. Not surprisingly, we observed substantial differences in carbon pools and fluxes between these different cover types; meadows, aspen forests, old coniferous forests and young coniferous forests were all unique in some aspects of carbon dynamics. As a consequence, an *a priori* stratification into important cover types likely would have helped us ensure that our plots accurately represented the distribution of types across the landscape and thus facilitate more accurate large-scale estimates.

Our initial experiences suggest other lessons that may be specific to these high altitude southern Rocky Mountain forests. We found that carbon stored in both forest floor and coarse woody debris was very substantial, variable across the landscape, and not particularly well related to stand structure or forest age. This implies that accounting for these pools is both necessary and potentially challenging. Our initial impressions of cost and benefit for various measurements are illustrated in Fig. 17.1. In addition, we appreciated the grid-based plot selection because it challenged our bias about the structure and composition of "typical" forests in this region.

17.7 Conclusions

Accurately assessing landscape-scale terrestrial carbon dynamics with field measurements is a daunting, yet necessary, task. By critically evaluating our overall approach, we hope to identify potential areas for improvement and thereby strengthen future efforts. Although our project is ongoing, our initial experience establishing plots and conducting analyses suggests five primary lessons:

Standardization facilitates comparison – The consistent plot layout and sampling strategy that we employed enabled very straightforward comparisons between sites that are only beginning to be exploited.

Stratify the landscape – We found that many of our sites were diverse enough to warrant stratification prior to establishing plots, and that this stratification would likely have led to more efficient plot selection and more accurate overall carbon estimates. Ideally, this process would identify the sources of variability in carbon pools and fluxes, stratify across those sources, establish plots within each class, compare processes within and between classes to ensure the validity of the classification, and use the stratification to generate landscape-scale estimates.

Carefully consider plot layout – Plot layout must efficiently represent the landscape while minimizing bias. The FIA plot design is not efficient for characterizing short-term processes and changes – fewer subplots/plot would achieve similar accuracy with less effort.

Evaluate sampling design and intensity for each measurement – Researchers should consider how variability in space and time influences sampling intensity and frequency and balance these requirements against the impact of the process on total carbon storage and balance. A one-size fits all sampling strategy is probably not best for all sites, and certainly not most efficient for all variables at any site. Keep in mind that measuring all variables at all plots facilitates good comparison between variables, but is likely not the most efficient approach.

Anticipate compatibility with other approaches – Consider conducting measurements in a way that will provide robust ground data for imagery at smaller scales, where we have confidence in our estimates. This finer-resolution imagery can be used to scale up to the whole square kilometer. Work with models/modelers prior to conducting measurements to ensure that comparable processes are being measured.

Literature Cited

Baldocchi, DD (2003) Assessing the eddy covariance technique for evaluating carbon dioxide exchange rates of ecosystems: past, present and future. Global Change Biology 9:479–492

Barford, CC, Wofsy, SC, Goulden, ML, Munger, JW, Pyle, EH, Urbanski, SP, Hutyra, L, Saleska, SR, Fitzjarrald, D, Moore, K (2001) Factors controlling long- and short-term sequestration of atmospheric CO_2 in a mid-latitude forest. Science 294:1688–1691

Bechtold, WA, Patterson, PL (2005) The enhanced forest inventory and analysis national sample design and estimation procedures. SRS-80, U.S. Department of Agriculture, Forest Service, Southern Research Station, Asheville, NC

Bradford, JB, Weishampel, PA, Smith, ML, Kolka, RK, Ryan, MG, Birdsey, RA (2008) Carbon pools and fluxes in temperate forest landscapes: spatial variability and implications for sampling design. Landscape Ecology (in review)

Chen, JQ, Franklin, JF, Spies, TA (1992) Vegetation responses to edge environments in old-growth Douglas-fir forests. Ecological Applications 2:387–396

Cochran, WG (1977) Sampling techniques. Third edition. Wiley, New York

Curtis, PS, Hanson, PJ, Bolstad, P, Barford, C, Randolph, JC, Schmid, HP, Wilson, KB (2002) Biometric and eddy-covariance based estimates of annual carbon storage in five eastern North American deciduous forests. Agricultural and Forest Meteorology 113:3–19

Euskirchen, ES, Harper, KA, Qinglin, L (2006) The influence of edges on plant communities. In: Chen, GQ, Saunder, SC, Brosofske, KD, Crow, TR (Eds.) Ecology of hierarchical landscapes - from theory to application. Nova Science Publishers, Inc, New York, pages 71–88

Grunblatt, J (1987) An MTF analysis of Landsat classification error at field boundaries. Photogrammetric Engineering and Remote Sensing 53:639–643

Hansen, AJ, Rotella, JJ, Kraska, MPV, Brown, D (2000) Spatial patterns of primary productivity in the greater Yellowstone ecosystem. Landscape Ecology 15:505–522

Noormets, A, Ewers, B, Sun, G, Mackay, S, Zheng, D, McNulty, S, Chen, GQ (2006) Water and carbon cycles in heterogeneous landscapes: an ecosystem perspective. In: Chen, GQ, Saunder, SC, Brosofske, KD, Crow, TR (Eds.) Ecology of hierarchical landscapes - from theory to application. Nova Science Publishers, Inc., New York, pages 89–124

Rossi, RE, Mulla, DJ, Journel, AG, Franz, EH (1992) Geostatistical tools for modeling and interpreting ecological spatial dependence. Ecological Monographs 62:277–314

Ryan, MG, Law, BE (2005) Interpreting, measuring, and modeling soil respiration. Biogeochemistry 73:3–27

Turner, DP, Ollinger, SV, Kimball, JS (2004) Integrating remote sensing and ecosystem process models for landscape- to regional-scale analysis of the carbon cycle. Bioscience 54:573–584

Turner, DP, Ritts, WD, Cohen, WB, Maeirsperger, TK, Gower, ST, Kirschbaum, AA, Running, SW, Zhao, MS, Wofsy, SC, Dunn, AL, Law, BE, Campbell, JL, Oechel, WC, Kwon, HJ, Meyers, TP, Small, EE, Kurc, SA, Gamon, JA (2005) Site-level evaluation of satellite-based global terrestrial gross primary production and net primary production monitoring. Global Change Biology 11:666–684

Wagner, HH, Fortin, MJ (2005) Spatial analysis of landscapes: concepts and statistics. Ecology 86:1975–1987

Zheng, D, Hunt, ER, Doraiswamy, PC, McCarty, GW, Ryu, S (2006) Using remote sensing and models to understand the ecology of landscapes. In: Chen, GQ, Saunder, SC, Brosofske, KD, Crow, TR (Eds.) Ecology of hierarchical landscapes - from theory to application. Nova Science Publishers, Inc., New York, pages 125–166

Index

M

N

P

R

S

T

V

W